J. WESTON WALCH PUBLISHER
Portland, Maine

EASY
Science Demos & Labs

Physics

Thomas Kardos

User's Guide
to
Walch Reproducible Books

Purchasers of this book are granted the right to reproduce all pages.

This permission is limited to a single teacher, for classroom use only.

Any questions regarding this policy or requests to purchase further reproduction rights should be addressed to

Permissions Editor
J. Weston Walch, Publisher
321 Valley Street • P.O. Box 658
Portland, Maine 04104-0658

1   2   3   4   5   6   7   8   9   10
ISBN 0-8251-4502-3

Contents

Preface . . . . *vii*

National Science Education Standards . . . . *viii*

Suggestions for Teachers . . . . *x*

Equipment . . . . *xi*

Safety Procedures . . . . *xii*

Demos and Labs

DEMO 1 Physical Change and Properties of Matter 3

DEMO 2 Energy Waves and Energy Forms 6

DEMO 3 Energy Has Different Properties Than Matter 9

DEMO 4 Energy (Heat) Expands Matter 10

DEMO 5 Absorption of Heat 12

DEMO 6 Radiant Energy 13

DEMO 7 Vacuum Bottles 14

DEMO 8 Kinetic Molecular Theory: States of Matter 15

DEMO 9 Pressure of Air 17

DEMO 10 Air Pressure and Soda Can 18

DEMO 11 Work from Air Pressure 19

DEMO 12 Light Travels in a Straight Line 20

DEMO 13 Pinhole Camera 21

DEMO 14 Angle of Incidence Equals Angle of Reflection 22

STUDENT LAB 1 Fermat's Principle of Least Time 23

DEMO 15 Mirror Images: Depth 25

DEMO 16 Bending Light: Laws of Refraction 26

DEMO 17 Motion Picture Effect 27

DEMO 18 Electromagnetic Spectrum 28

DEMO 19 Frequency, Wavelength, and Amplitude 29

DEMO 20 The Light Spectrum: Color of Objects 30

DEMO 21 Laser Light 31

DEMO 22 Lenses 33

DEMO 23 Refraction of Light #1 34

DEMO 24 Refraction of Light #2 35

STUDENT LAB 2 Young's Two Slit Experiment 36

DEMO 25 Breaking Light Apart 37

STUDENT LAB 3 Total Internal Reflection 39

DEMO 26 Potential Energy #1 40

DEMO 27 Potential Energy #2 41

DEMO 28 Speed, Velocity, and Friction 42

STUDENT LAB 4 Constant Velocity 43

DEMO 29 Acceleration 45

STUDENT LAB 5 Law of Falling Bodies (Galilean Accelerator) 47

STUDENT LAB 6 Momentum 49

DEMO 30 Bernoulli's Principle 51

DEMO 31 Newton's Third Law of Motion 53

DEMO 32 Magnets and Poles 54

DEMO 33 Making a Magnet 55

DEMO 34 Inducing Magnetism 56

DEMO 35 Magnetic Fields 57

DEMO 36 Magnetic Compass 58

DEMO 37 Destroying a Magnet 59

DEMO 38 Static Electricity 60

DEMO 39 Static Electricity: Van de Graaff Generator 61

DEMO 40 Electricity Makes a Magnet 63

DEMO 41 Electromagnets 64

DEMO 42 Making Electromagnets Stronger 65

DEMO 43 Inducing Electricity by Magnetism 66

DEMO 44 Inducing Electricity: Electromagnets 68

STUDENT LAB 7 Building a Simple Circuit 69

DEMO 45 Series Circuits 71

DEMO 46 Parallel Circuits 72

DEMO 47 Short Circuits 73

DEMO 48 Measuring Voltage 74

DEMO 49 Measuring Current 75

DEMO 50 Measuring Resistance 76

DEMO 51 Ohm's Law 77

DEMO 52 Thermometers and Temperature 79

DEMO 53 Calories/Calorimeter 80

STUDENT LAB 8 Second Law of Thermodynamics 81

STUDENT LAB 9 Entropy 82

DEMO 54 Force: Measuring Force 83

DEMO 55 Work: Measuring Work 84

DEMO 56 Measuring Friction 85

DEMO 57 Center of Gravity 86

DEMO 58 Lever: Mechanical Advantage 87

DEMO 59 Wheel and Axle 89

DEMO 60 Teeter-Totter: Moment 90

DEMO 61 The Inclined Plane 91

DEMO 62 Pulleys 92

DEMO 63 Friction and Machines 93

DEMO 64 Gravity 94

DEMO 65 Acceleration by Gravity 96

DEMO 66 Weight 97

STUDENT LAB 10 Calculating Special Relativistic Effects 98

DEMO 67 Pressure 101

DEMO 68 Measuring Pressure 102

DEMO 69 Density and Specific Gravity 104

DEMO 70 Archimedes' Principle 106

DEMO 71 Making Sounds 107

DEMO 72 Loudness and Sound (Decibels and Amplitude) 108

DEMO 73 Pitch (Frequency) and Sound 110

DEMO 74 Transmission of Sound Through Materials 111

DEMO 75 Resonance 112

Appendix

1. Assessing Laboratory Reports 115

2. Temperature Conversion (Celsius to Fahrenheit) 117

3. Temperature Conversion (Fahrenheit to Celsius) 118

4. Melting and Boiling Points of Elements 119

5. Range of Resistances 120

6. Coefficients of Volume Expansion 120

7. Electrochemical Equivalents 120

8. Wavelengths of Various Radiations 120

9. Specific Heat of Materials 121

10. Coefficient of Linear Expansion 121

11. Density of Liquids 122

12. Altitude, Barometer, and Boiling Point 122

13. Specific Gravity 122

14. Units: Conversions and Constants 123

Glossary 140

Preface

As a middle school teacher, many times I found myself wishing for a quick and easy demonstration to illustrate a word, a concept, or a principle in science. Also, I often wanted a brief explanation to conveniently review basics and additional information without going to several texts.

This book is a collection of many classroom demonstrations. Explanation is provided so that you can quickly review key concepts. Basic science ideas are hard to present on a concrete level; the demonstrations fill that specific need. You will also find 10 specially created laboratory activities for middle school students that are safe enough for young people to do on their own. These labs add a deeper level of understanding to the demonstrations.

An actual teacher demonstration is something full of joy and expectation, like a thriller with a twist ending. Keep it that way and enjoy it! Try everything beforehand.

We need to support each other and leave footprints in the sands of time. Teaching is a living art. Happy journey! Happy sciencing!

—*Thomas Kardos*

National Science Education Standards for Middle School

The goals for school science that underlie the National Science Education Standards are to educate students who are able to

- experience the richness and excitement of knowing about and understanding the natural world;

- use appropriate scientific processes and principles in making personal decisions;

- engage intelligently in public discourse and debate about matters of scientific and technological concern; and

- increase their economic productivity through the use of the knowledge, understanding, and skills of the scientifically literate person in their careers.

These abilities define a scientifically literate society. The standards for content define what the scientifically literate person should know, understand, and be able to do after 13 years of school science. Laboratory science is an important part of high school science, and to that end we have included student labs in this series.

Between grades 5 and 8, students move away from simple observation of the natural world and toward inquiry-based methodology. Mathematics in science becomes an important tool. Below are the major topics students will explore in each subject.

- Earth and Space Science: Structure of the earth system, earth's history, and earth in the solar system

- Biology: Structure and function in living systems, reproduction and heredity, regulation and behavior, populations and ecosystems, and diversity and adaptations of organisms

- Chemistry and Physics: Properties and changes of properties in matter, motions and forces, and transfer of energy

Our series, *Easy Science Demos and Labs*, addresses not only the national standards, but also the underlying concepts that must be understood before the national standards issues can be fully explored. By observing demonstrations and attempting laboratory exercises on their own, students can more fully understand the process of an inquiry-based system. Cross-curricular instruction, especially in mathematics, is possible for many of these labs and demonstrations.

Suggestions for Teachers

1. A • (bullet) denotes a demonstration. Several headings have multiple demonstrations.

2. **Materials:** Provides an accurate list of materials needed. You can make substitutions and changes as you find appropriate.

3. Since many demonstrations will not be clearly visible from the back of the room, you will need to take this into account as part of your classroom management technique. Students need to see the entire procedure, step by step.

4. Some demonstrations require that students make observations over a short period of time. It is important that students observe the changes in progress. One choice is to videotape the event and replay it several times.

5. Some demonstrations can be enhanced by bottom illumination: Place the demonstration on an overhead projector and lower the mirror so that no image is projected overhead.

6. I use a 30-cup coffeepot in lieu of an electric hot plate, pans, and more cumbersome equipment to heat water for student experiments and to perform many demonstrations.

7. As the metric system is the proper unit of measurement in a science class, metric units are used throughout this book. Where practical, we also provide the U.S. conventional equivalent.

8. Just a few demonstrations may appear difficult to set up, for they have many parts. Be patient, follow the listing's steps, and you will really succeed with them.

Equipment

- Sometimes, though rarely, I will call for equipment that you may not have. An increasing growth in technology tends to complicate matters. Skip these few demonstrations or borrow the equipment from your local high school teacher. Review with him or her the proper and safe use of it. These special demonstrations will add immensely to your power as an effective educator and will enhance your professionalism.

- Try all demonstrations in advance to smooth your show. If something fails, enjoy it and teach with it. Many great scientific discoveries had to be done over many times before their first success. Edwin Land had to do more than 11,000 experiments to develop the instant color photograph. Most people would have quit long before that.

- One of my favorite techniques is to record with a camcorder and show the demonstration on a large monitor.

Safety Procedures

- Follow all local, state, and federal safety procedures. Protect your students and yourself from harm.

- Attend safety classes to be up-to-date on the latest in classroom safety procedures. Much new legislation has been adopted in the recent past.

- Have evacuation plans clearly posted, planned, and actually tested.

- Conduct experiments involving chemicals only in rooms that are properly ventilated.

- Have an ABC-rated fire extinguisher on hand at all times. Use a Halon™ gas extinguisher for electronic equipment.

- Learn how to use a fire extinguisher properly.

- Label all containers and use original containers. Dispose of chemicals that are outdated.

- Know and teach an adequate method for disposing of broken glass.

- Heat sources, such as Bunsen burners and candles, can be hazardous. Use caution when heating chemicals.

- Wear required safety equipment at all times, including goggles, gloves, and smocks or labcoats.

- Never eat or drink in the laboratory.

- Practice your demonstration if it is totally new to you. A few demonstrations do require some prior practice.

- Conduct demonstrations at a distance so that no one is harmed should anything go wrong.

- Have students wash their hands whenever they come into contact with anything that may be remotely harmful to them, even if years later, like lead.

- Neutralize all acids and bases prior to disposal, if possible, in a chemical fume hood.

- Dispose of demonstration materials in a safe way. Obtain your district's guidelines on this matter.

- Be especially aware of the need to dispose of hazardous materials safely. Some chemistry experiments create byproducts that are harmful to the environment.

- Take appropriate precautions when working with electricity. Make sure hands are dry and clean, and never touch live wires, even if connected only to a battery. Never test a battery by mouth.

- When using lasers, never look directly into the beam, and make sure students understand the dangers of laser light.

Disclaimer: The safety rules are provided only as a guide. They are neither complete nor totally inclusive. The publisher and the author do not assume any responsibility for actions or consequences in following instructions provided in this book.

Demos and Labs

A **physical change** occurs when a material alters in size or shape but still remains the same material.

Materials: piece of paper, a rubber band, a small piece of wood (splint)

- Take a piece of paper and tear it. It is still paper. Take a rubber band and stretch it. Take a small piece of wood and break it.

Material objects have many physical **properties.** Some properties are listed here with examples of attributes that describe them. Your students may benefit by using this list when describing objects. By repeating this activity many times in a year, students become better observers. This section is provided only as a quick reference.

TABLE OF PROPERTIES

Property	Example
Color	Blue, red, green
State	Solid, liquid, gas, (plasma)
Measurements	Length, width, height, mass, volume
Material	Paper, wood, metal, plastic
Temperature	How hot is it?
Quantity	How many are there?
Transparency	Can you see through it?
Magnetism	Does a magnet attract it?
Electrical conductivity	Conductor, semiconductor, or insulator
Buoyancy	Does it float?
Sound	Does it make noises on standing?
Edibility	Can you eat it?
Odor	Does it smell?
Fragility	Does it break easily?

(continued)

TABLE OF PROPERTIES *(continued)*

Property	Example
Hollowness	Is it empty inside?
Fluorescence	Does it glow under black (UV) light?
Layers	Is it made in layers?
Moisture	Is it wet? dry?
Flexibility	Can you bend it?
Shine	Does it have a luster, or is it dull?
Hardness	Firm, solid, soft, spongy
Texture	Rough, bumpy, smooth
Holes	Does it have holes? where? how many?
Solvency	Does it dissolve materials? what kinds of materials?
Solubility	Does it become dissolved when placed in liquids?
Acid material	Check with litmus paper: both red.
Base material	Check with litmus paper: both blue.
Neutral material	Neither litmus changes colors.
Schlieren	Does it produce stringers of coloring (schlieren)?

- Display an array of objects and describe them in terms of one or several of these properties.

(continued)

On a higher level, some physical properties can be described as follows:

TABLE: PHYSICAL PROPERTIES OF MATTER

Physical Property	Unit
Density	10^3 g cm^{-3}
Young's modulus	10^{10} N m^{-2}
Specific heat capacity	$\text{J kg}^{-1} \text{ K}^{-1}$
Specific latent heat of fusion	10^4 J kg^{-1}
Thermal conductivity	$\text{W m}^{-1} \text{ K}^{-1}$
Resistivity	$10^{-8}\rho \text{ m}$
Linear expansivity	10^{-6} K^{-1}

Density: mass per unit of volume. Without units, it is specific gravity, with water set as 1. Objects with S.G. < 1 float; objects with S.G. > 1 sink.

Young's modulus: deals with the elasticity of a material and how much it can be stretched before it deforms

Specific heat: deals with the speed at which heat energy can be absorbed and given off

Heat of fusion: deals with the amount of heat energy needed to change the state of the material (e.g., solid to liquid), different than in just normal heating. Each material has its own heat of fusion.

Thermal conductivity: the amount of heat energy that a specific substance can conduct

Resistivity: the resistance to conducting an electric current, unique to each material and low in most metals

Linear expansivity: the amount of linear expansion when heat is applied

Note: These variables change with temperature.

Energy travels in waves. Waves are either transverse or longitudinal. **Transverse waves** vibrate (oscillate) at right angles to the direction of the wave. When two waves travel in opposite directions, they "buck" or oppose each other. When two waves travel in the same direction, they "aid" each other, or add their energies together.

Materials: about 10 to 15 feet of rope or string, or a Slinky™ (The metal ones work best.)

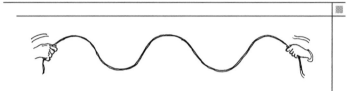

- Have two people hold a rope about 8 to 12 feet apart, and ask one of them to swing the rope gently up and down on one end. Observe how the energy of small swings is transmitted to the other end in a wavelike motion.

- Have two people take a Slinky and pull it apart about 15 feet. Let them wave it up and down gently. Observe the energy waves. Have them start a wave on each end simultaneously, so the class can observe what happens when two waves oppose (buck) each other. Ask one volunteer to send several waves away from the same end, showing how waves help (aid) each other.

Longitudinal waves vibrate (oscillate) in the direction of the wave. An example is sound waves.

Materials: Slinky (The metal ones work best.)

- Stretch a Slinky about 15 to 20 feet. Have a student hold the other end securely. Wait till the spring stops moving. Carefully squeeze two or three spring loops together and watch the "squeeze" travel forth and back along Slinky. These are longitudinal waves. In the places where the springs are closer than normal, you have wave compression. In the places where the springs are fully apart, you have wave rarefaction.

(continued)

- While holding the Slinky ready, as in the previous demonstration, touch the end of it to a student's ear. Generate several longitudinal waves. The student will be able to hear the vibrations of the spring. The vibrations will be similar to the sounds of a synthesizer.

Following are some forms of energy:

TABLE: ENERGY FORMS

Form	Explanation
Sound	Waves in air caused by vibrations, such as horn blowing, objects hitting, talk, bark of dog, etc.
Electrical	Natural and man-made energy form; travels in wires; convenient to use
Chemical	Stored in plants, animals, and minerals; digestion releases food energy
Atomic	Energy stored in the nucleus of the atom
Light	A form of energy produced by the sun and lamps; visible to the human eye
Heat	Causes molecules to move; produced by human digestion and changes from other energy forms
Mechanical	The energy in the motion of objects, such as moving machines, falling objects, etc.

Materials: small light with bulb, flashlight, baking soda, vinegar, glass with some water, teaspoon, keys on a ring

- Turn the small light on. Observe the light energy. Mention that after a few minutes the bulb will be too hot to touch. This invisible radiant energy is heat energy (infrared).

- Place a couple of teaspoons or a tablespoon of baking soda in a glass half-filled with water and stir it. Add some vinegar to this mixture. A fairly quick and violent chemical reaction will result. (Perform this demonstration over a sink.)

(continued)

- Turn on the flashlight. The chemical energy from the reaction in the battery makes electricity for the flashlight.

- Make a sound by jingling your keys or tapping your desk with another object. The vibrations you make will result in sound energy that can be heard.

These demonstrations will clarify that, while interchangeable, energy and matter have very different properties.

Materials: flashlight, scale, glass full of water

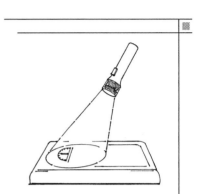

- Take a flashlight and shine it on a bathroom scale. Notice that when the beam strikes the scale, the scale shows no increase in weight. Energy has no mass.

- Shine a flashlight into a glass filled to the brim with water. The water does not overflow. Energy takes up no space.

- Energy keeps molecules (tiny bits of matter) in continuous motion. Fill a glass with water. Let it stand 5 to 10 minutes until the water is still. Place one drop of food coloring near the top of the water and let students observe for 5 or 6 minutes the mixing of the water with the coloring. (Lighting it from below will accent the color mixing.) The molecules of water and color strike each other due to their normal **Brownian motion,** and in the process they mix together. The stringers of coloring are schlieren.

When energy in the form of heat is added to solids, liquids, and gases, matter expands. This occurs because molecules contained within the matter are excited by the heat, and move more. Given enough application of heat, solids and liquids can change their states.

Materials: for solids—brass ring with a tight-fitting brass ball (obtainable from scientific supply house, or make your own apparatus—see next page), Bunsen burner

for liquids—small bottle, cork to fit bottle, soda straw, food coloring, water, florist's clay, pan with water, stove or electric coffeepot

for gases—small bottle, balloon, pan with water, stove or electric coffeepot

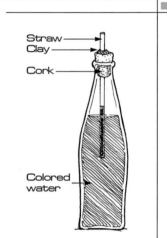

- Solids: Show students how the brass ball moves freely through the brass ring at room temperature. Heat the ball until it does not go through the ring. The brass has expanded. Cool the brass ball by dipping it in water. Show how it goes through the ring again.

- Liquids: Fill a bottle about 3/4 full with colored water. Make a hole in the cork and insert the soda straw until the straw is about 1 inch above cork. Place the cork in the bottle and seal the soda straw with clay. Notice the height of the water in the soda straw. Place the bottle in the pan full of hot water or in a coffeepot. Notice how the water level in the straw rises. The water has expanded.

(continued)

- Gases: Place a balloon on top of a small bottle. Place the bottle in hot water. Notice how the balloon inflates. The air inside the bottle and balloon has expanded due to the heat energy. Place the bottle in cold water and notice how the balloon contracts.

Following are directions for making your own apparatus for demonstrating the expansion of solids.

Materials: two dowels, one screw, one eye screw (with opening to fit the screw's diameter)

- Screw the screw 3/4 of the way into the end of one dowel. Do the same with the eye screw and the other dowel. Use these dowels in place of the brass ring and ball in the demonstration for the expansion of solids.

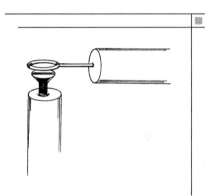

Heat energy, nonvisible **infrared** rays from the sun or from other luminous sources, travels and behaves similarly to visible light. Most shiny and smooth surfaces reflect heat, most dark and rough ones absorb it, and most transparent ones let it go right through. In sunshine, chromed car bumpers are cooler than black ones. People in the tropics wear light-colored clothing, while Arctic explorers wear dark clothes. Houses in warm climates are white to reflect heat. Low-power infrared remotes are used by the home consumer industry for controlling TVs, VCRs, etc.

Materials: two identical bottles with narrow necks, black tempera paint, two balloons, two same-sized empty tin cans, cardboard, water, two thermometers

Plain bottle   Painted bottle

- Paint one bottle black. Fill the two bottles with water to the same level. Stretch the balloons to check for their expansion. Snap a balloon on each bottle. Let the bottles stay in the sun for a while. The painted bottle will inflate the balloon more; its air has expanded more because the black paint absorbs more heat energy. The shiny glass has allowed the infrared rays to pass through or it has reflected them.

- Paint one tin can black. Place the same amount of water in both cans. Cover both tin cans with a piece of cardboard. Pierce a hole in each piece and insert the thermometers. Place both cans in the sun for 15 to 20 minutes. Again, the black can will show a higher water temperature than the shiny one.

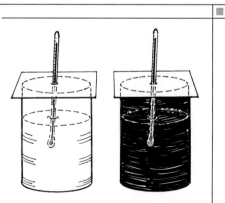

Heat **radiates** out from a substance toward a region of less heat. Heat energy goes through a transparent substance, is reflected by a smooth and shiny surface, and is absorbed by a rough or dark substance.

Materials: two identical shiny tin cans, black tempera paint, hot water, two thermometers, cardboard

- Paint one tin can with black tempera paint. Fill both cans with the same amount of hot water. Cover each with cardboard. Pierce a hole in the cardboard and insert the thermometers. Let the cans sit in a shady place. After 15 to 20 minutes, let your students note that the black can has radiated more heat, for its thermometer shows a lower temperature.

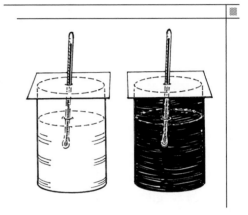

Vacuum bottles are an application of the principles of heat reflection. The inner part of the container is a double-walled glass or metal jar, silvered on the inside and on the outside. The space in between the glass walls is a vacuum; most of its air molecules were pumped out. This empty space is a poor conductor of heat. The double walls and the space in between provide three barriers to heat transmission. If you place hot liquids in the bottle, the heat is reflected back. If you place cold liquids in the bottle, heat is radiated away from the bottle, keeping it from entering. In this manner, liquids can be kept at the desired temperature.

Materials: vacuum bottle (thermos), flashlight

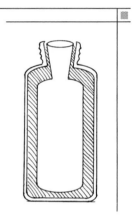

- Bring to school an empty vacuum bottle and let your students look at it. If possible, unscrew the top and let them see the inner bottle by itself. Handle with care, for glass jars are fragile. Shine the flashlight inside the bottle to show the silvering.

The kinetic molecular theory explains how matter behaves and how it changes from one state into another. Molecules in **solids** are held quite closely together by molecular bonds. All molecules vibrate. If heat energy is added, the molecules move faster and farther apart, until they begin to slide over each other. At this point, the solid has changed into a **liquid.** If more energy is added, some molecules escape from the surface of this loose bond in the liquid and begin to move even farther apart. The new state of these escaping molecules is **gas.** If energy is removed, a gas changes into a liquid and a liquid changes into a solid. These are the phases (states) of matter.

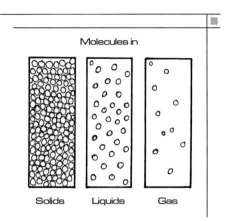

Molecules in

Solids   Liquids   Gas

Materials: ice cubes, ice cube tray, water, freezer

- Take several ice cubes and let them melt in a glass until they change into water. Ice cubes melt, for heat energy *is added.*

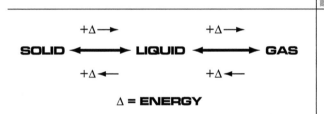

$$SOLID \xrightarrow{+\Delta} LIQUID \xrightarrow{+\Delta} GAS$$

$$\Delta = ENERGY$$

- Fill an ice cube tray with water and place in a freezer. Water, a liquid, becomes a solid. Heat energy *has been removed.* Changes of states are reversible.

Materials: three identical 10-inch-long boxes, clear plastic wrap, masking tape, 15 marbles

- Remove the tops of the boxes and modify two of them to be 5 inches and 7 inches long. The boxes will represent molecules in solid, liquid, and gaseous states. Place 5 marbles in each and cover them with transparent plastic wrap. Tape the wrap. Have students shake the 5-inch box. Then have them shake the 7-inch box a bit harder and the 10-inch box even harder. The behavior of the marbles resembles the behavior of molecules in matter.

(continued)

Materials: coffeepot, water

- Take a coffeepot, fill it partially with water, and let it warm. Make certain that its lid is in place. After the water has warmed up (in electric units a warning light goes on or off), lift the lid and let students observe how water is forming drops on the top of the lid. Have them observe the whitish cloud coming up from the pot. They will call it steam, but it is not. Steam is invisible. What they are seeing is the water vapor that forms as the invisible steam cools down and condenses into many tiny water droplets that look white. The lid, a little bit cooler than the air inside the coffeepot, also condenses gaseous water and forms water droplets. This is why water drips off the lid when you uncover the coffeepot or any other cooking pot.

Materials: a glass or two, several ice cubes, insulated container

- Place a glass in an insulated container and place several ice cubes around the glass. Let the glass become chilled. Take the glass out and show the class how it frosts. This frost is the **condensation** of water in the air. The molecules of water condense since the glass is colder than the environment.

When energy is unbalanced—i.e., higher versus lower pressure, whether it is osmotic, electrical (voltage), or physical (air pressure)— energy flows from the region of higher pressure to the lower one until a balance (**equilibrium**) is reached. If a space capsule loses its pressure integrity, life-supporting air escapes out to the vacuum of space, with fatal consequences for the spacefarers. On Earth, the pressure of the atmosphere is 14.7 lb/in^2. This pressure is sufficient to crush any container with a vacuum inside it, unless the container is specifically constructed to withstand these forces.

Materials: hard-boiled egg (peeled), birthday candle, matches, large bottle or jar with a mouth barely smaller than the egg

- Stick the candle into the egg on one of its ends and light it. Place the candle inside the bottle and seal the opening with the egg. As the candle burns the oxygen—20% of the inside air—the bottle will have a region of lower pressure. Eventually the egg will squeeze into the bottle. (During this demonstration, be careful to position your hands so that students can see how things happen.)

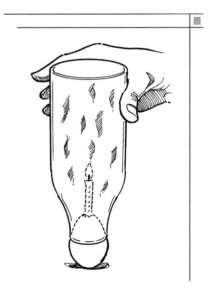

- To remove the egg, angle the bottle opening down and blow air into it. Then point the opening straight down. The air around the egg will have more pressure and the egg will squeeze out.

Materials: empty soda can, tablespoon, small pan with water, forceps, Bunsen burner or hot plate

- Place a tablespoon (15 mL) of water into an empty soda can. Heat it on a Bunsen burner with forceps until steam begins to escape from the pop-top. Rapidly invert the can and place it in the water bath. The cooling of the air inside the can will create enough difference in air pressure that the can will be crushed by the atmospheric pressure. This happens because the steam escaping the can leaves fewer molecules. By quickly sealing the opening (cool matter will not rise into warm), you have created a relative vacuum inside the can.

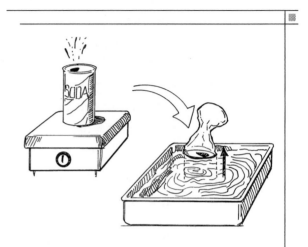

The pressure of air can be made to do useful work. One example is the aerosol bottles used for perfumes, deodorants, paints, and so on. Divers use tanks of compressed air. There are many compressed-air (pneumatic) tools in workshops: air hammers, air buffers, air ratchets, air drills, and so on. Compressed air is used in tires, footballs, basketballs, air brakes, riveters, and sandblasters.

Materials: bottle, cork with hole, eyedropper, small piece of rubber hose to fit dropper, water

- Insert the dropper in the cork so that the narrow part is up, leaving a small portion extending from the bottom side. Connect the rubber hose to the bottom of the dropper. Cut the hose just long enough to reach to the bottom of the bottle. Half-fill bottle with water and close the bottle with the cork. Blow vigorously into the bottle through the eyedropper. Within seconds, the bottle will become a fountain. The air fills the space above the water and becomes compressed. The pressure pushes the water out. This is similar to how aerosol bottles work.

Light travels in a straight line. This experiment will demonstrate the principle that light propagates—transmits its energy—in a straight line.

Materials: several index cards, small pieces of clay, flashlight, hole punch, soda straw(s)

- Punch a small hole in the center of each card. (Draw diagonal lines to find the center.) Stand all cards vertically by pushing each one into a blob of clay. Place the flashlight at one end and arrange the cards so that light can be seen through all the holes. The holes must be arranged in a straight line if light is to go through them. Experiment with finer holes.

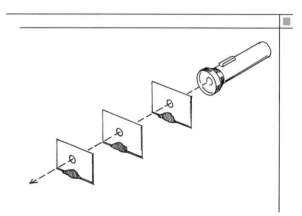

- Have your students look through a soda straw at the flashlight in a darkened room. Have each one bend the straw. They will notice that no light will go through it.

(continued)

A pinhole camera is another way to demonstrate that light travels in a straight line.

Materials: one small cardboard box, small piece of aluminum foil, wax or tissue paper, adhesive tape, needle, flat black paint

- Take the following steps:

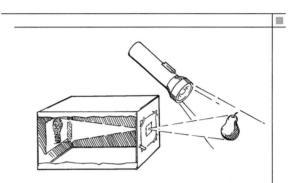

  1. Cut a hole about 1 inch square in one end of the box. Make certain that it is in the center. (Drawing diagonal lines will help locate the center.)

  2. In the opposite end of the box, cut an opening at least 4 inches square. Try to have it centered.

  3. Cover the smaller hole with flat aluminum foil, and tape the edges of the foil flat to the box with the adhesive tape.

  4. Paint the inside of the box with flat black paint and let it dry.

  5. With the needle, pierce a small hole in the center of the aluminum foil.

  6. Cover the larger hole with tissue or wax paper.

  7. Place a small object in front of the pinhole camera and light it with the flashlight. Make certain that the flashlight is above the camera.

  8. You will be able to observe the object on the paper in the back of the pinhole camera. The image will be inverted. This is additional proof that light travels in a straight line. (Notice the two intersecting straight lines in the diagram.)

- Move the camera closer and farther away from the object. Show your students what happens to the size of the image in the back of the pinhole camera.

A key property of light is that when it strikes a reflecting or shiny surface, its angle of incidence (strike) is the same measure as its angle of reflection. A normal line is a line perpendicular (at 90°) to the reflecting surface. If the surface is smooth, like a mirror or a very calm body of water, there is a reflected regular image. If the reflecting surface is rough or the water is moving, then the image is diffused and no image can be seen. When photographing people or objects in front of shiny surfaces like mirrors or glass, photographers take the shot at an angle to the surface to avoid their own reflections and that of their flashes. Photographers also use high flash brackets to avoid having people with red eyes in photographs.

Materials: flashlight, black paper, mirror, clay, masking tape, razor blade, protractor, small ball, straight edge (such as a meterstick), pencil, large sheet of paper

- Before you do the next demonstration, bounce a ball on the floor at an angle. It will bounce back at a similar angle. Discuss with your students that people who play pool and billiards are very careful about the angle at which they hit the balls. If the billiard balls hit the table's side at an angle they will bounce back at the same angle, unless they have some spin (English) applied to them.

- Cover the flashlight with black paper and cut a narrow slit in the paper. Place the mirror vertically by sticking it into a blob of clay. On a larger piece of paper, draw a long, straight line *AB*. Somewhere in the middle, mark a point C and draw a line at 90° to C. Label it *CD*. This 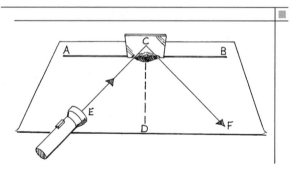 is a normal line to line *AB* at point C. Place the mirror's back surface parallel and on top of line *AB*, centered on point C. Keeping the slit vertical, shine the flashlight on its narrow beam at point C. Mark both the ray of incidence (striking) light and the ray of reflected light. With a protractor, measure the angles of incidence and reflection. They should be the same. (If they aren't, the mirror was not vertical.)

You know that light normally travels in straight lines. In going from one place to another, light will take the most efficient path and travel in a straight line. This is true if there is nothing to obstruct the passage of light between the source and the places under consideration. If light is reflected by a mirror, the change from the otherwise straight path is a fairly simple formula. We must also consider the law of reflection, which is that the angle of incidence—the angle at which the light strikes the mirror—must equal the angle of reflection. In any case, light will take the shortest possible time to get from one place to another. This principle is called **Fermat's Principle of Least Time,** described by Pierre Fermat in 1650.

Materials: large plane mirror that can be held up on its side, laser pointer, some sort of mount for the laser pointer (could be a large ball of clay), protractor, large sheet of paper, pen, string

Procedure:

1. Lay the paper flat on a large surface, and arrange the mirror on its side in the middle.

2. Mark an arbitrary spot A on the positive side of the mirror (the side facing the reflective glass). Place the laser pointer on that dot, facing the mirror, but do not turn it on.

3. On the same side as the letter A, mark any arbitrary point B.

4. Use the string to follow that path from A across the mirror to the spot on the other side that is the same distance from the mirror as B. Mark that spot B'(B prime).

5. Using the string and protractor, ensure that B and B' are indeed crossing the mirror at a right angle.

6. Now, turn on the laser. The light will travel down to the mirror, and bounce back up to B. This will be easier to see if you darken the room slightly.

(continued)

7. Turn off the laser. Measure the angles of incidence and reflection, and note that they are the same.

8. Try it several times with several different *A* and *B* points.

Conclusion: Why does the principle of least time hold in all situations? Can you conceive of any situation where it wouldn't hold? Why or why not?

Strong Safety Warning: While this type of laser is harmless in general, it can cause serious eye damage if you look directly into its rays. Make sure that neither you nor the other students working in the vicinity are in the line of the lasers.

An image in a mirror appears reversed. The *real distance* of an object in front of a mirror is its distance from the mirror. *Depth distance* is the distance of a virtual image "behind" the mirror. The two distances are identical. If you want to photograph a person in a mirror, you have to adjust the camera focus for twice the distance between that person and the mirror. In flash photography, this depth distance becomes critical. If you are 10 feet from a mirror and so is your subject, you have to set the flash for 20 feet.

Materials: small mirror, two books, pencil, ruler, masking tape, construction paper about 20 × 20 inches

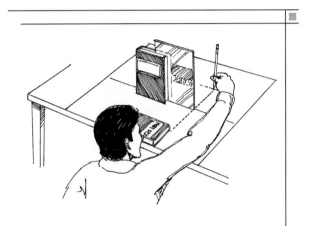

- Place a book or some written material in front of the mirror. In the mirror, it appears in reverse.

- Take a piece of construction paper about 20 × 20 inches, and place it on a table with one edge at the table's edge. Tape it in place. Draw a line across the middle of the paper, parallel to the table's edge. This is your reference line. Place the mirror in a vertical position by wedging it in a book. Place the mirror assembly on top of the reference line. Make certain that the back edge of the mirror is on the reference line. Place a book in front of the mirror with its written edge facing the mirror, about 6 inches away. Place your eye near the edge of the table. Look with both eyes at the mirror and the image of the book's edge in it. Move a vertical pencil beside the mirror, on one side. Move the pencil back and forth until it lines up with the book's reflection in the mirror. Mark this spot with the pencil. Measure the distance between the pencil mark and the rear edge of the mirror. Measure also the distance of the book's edge to the mirror. The two measurements should be nearly identical. (If the mirror is not vertical, the measurements may differ.) Use the back edge of the mirror as a reference for all your measurements.

Rays of light change their direction when they go from one transparent medium to another. This change of direction is **refraction.**

1. If light moves at an angle from a less dense medium into a denser one (for instance, from air into glass), it bends toward the normal line.

2. If light moves at an angle from a denser medium to a less dense medium (from water to air), it bends away from the normal line.

3. If light moves along the normal line, like a beam hitting a windowpane on the perpendicular, it is not bent.

Materials: water, fish tank, sheet of glass to cover part of fish tank, flashlight, black paper, razor blade, masking tape

- Add water to the fish tank up to about 3/4 full. Cover one end of the tank with a sheet of glass. Prepare the flashlight by taping black paper over the light and cutting a narrow slit in the black paper. In a darkened room, shine the flashlight's narrow beam at an angle through the glass into the water. It will show refraction.

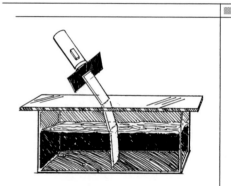

Motion pictures are possible due to a defect of the human eye. The eye will continue to see an image for 1/15 of a second after objects have been removed from view. This lasting image is called persistence of vision. By projecting still images rapidly, so that each new image comes to view prior to the extinction of the previous one, the eye sees images "moving." On television, the images are formed by scan lines, 525 per picture. The picture tube draws the odd and even lines alternately, 60 times a second, so that the persistence of vision makes it appear as a completed moving picture.

Materials: pencil with full eraser, razor blade, pin, index card

- Slit the pencil eraser in the middle carefully. Draw an image on each side of the index card. Insert the card into the slit and fasten with the pin. Twist the pencil rapidly with your hands, and both images will appear as one.

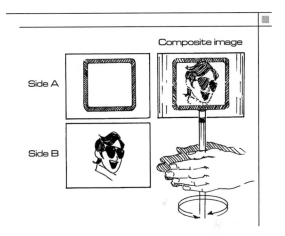

Side A

Side B

Composite image

Visible light is a small part of a larger spectrum of energy: the **electromagnetic spectrum.** Electromagnetic waves are transverse waves, which vibrate perpendicular to the direction of travel and consist of oscillating (vibrating) magnetic and electric fields. They have a wide range of frequencies and can travel through all media, including a vacuum. When they are absorbed, they cause a rise in temperature. These energies occur in random pulses (called **photons**), not in a continuous stream; they are explained by **quantum theory.**

TABLE: ELECTROMAGNETIC ENERGY

Cosmic rays	Background radiation; particles of enormous energy given off by stars
Gamma radiation	Deadly high energy given out by the sun and other stars
X rays	High energy used in X-ray machines
Ultraviolet rays	Invisible energy waves in sunlight, which cause skin to tan
Visible light	Basic colors of light, emitted by the sun; visible to the human eye
Infrared rays	Rays of heat energy; sensed by our nervous system
Radio waves	Microwaves; TV signals; radio energy

Materials: small lamp with bulb

- Show students a lightbulb in a lamp. Turn it on. They will immediately see the visible light, while they will have to feel the heat—an invisible form of energy. Explain that other forms of energy need special detectors.

Frequency (number of waves per second) varies inversely with **wavelength,** the width of a wave. As the number of waves (frequency) increases per unit of time, the wavelength becomes smaller. (See the diagram to the right. This is a key diagram to show your students, so they can understand this important relationship.)

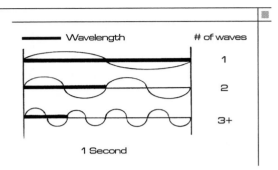

Frequency is measured in waves per second, called **hertz (Hz)** in honor of Heinrich Hertz, who did pioneering work in the field of electromagnetic waves.

Materials: cathode ray oscilloscope (CRO), sine-square wave generator, speaker, four to six wires with alligator clip ends

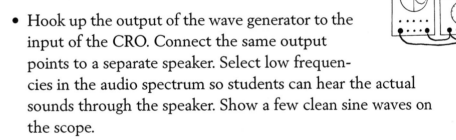

- Hook up the output of the wave generator to the input of the CRO. Connect the same output points to a separate speaker. Select low frequencies in the audio spectrum so students can hear the actual sounds through the speaker. Show a few clean sine waves on the scope.

- Increase the frequency of waves, and notice how the width of the individual waves becomes narrower to fit in the display. The waves have a shorter wavelength.

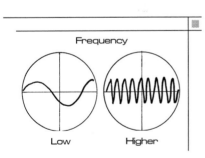

- Vary the output on the sine generator by changing its volume control. The sound will go from barely audible to loud. Observe the change in height of waves. The loud waves are higher than the soft ones. The height of the wave above or below the reference line is its **amplitude.** Amplitude measures the amount of force in the wave. For example, a tall ocean wave has more power than a small one.

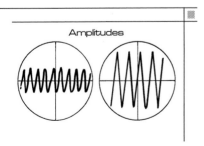

While most components of the electromagnetic spectrum are invisible, visible light can be seen. Light is a mixture of colors, and its color depends on its wavelength. Violet has the shortest wavelength, while red has the longest one. The other colors are in between these two. Ultraviolet light has shorter waves than violet light and is invisible. It causes skin to tan and it makes fluorescent objects glow. A longer wavelength than red is infrared. It is invisible heat energy, and one can only feel it. Objects reflect their own color and absorb all others. Black absorbs all colors, white reflects all colors, and red reflects only red. Transparent and translucent objects transmit their own color and block the rest. They act as color filters. If you shine a white light through a red filter, it will let only red light pass.

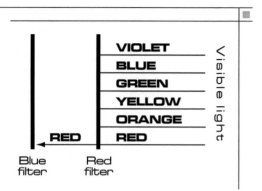

Materials: several color filters, colored pictures, pictures from magazines

- Look at the various colored pictures through the color filters. You will see only those components of the pictures that have the same color as the filter; the rest will be either white or invisible.

Easy Science Demos & Labs:
Physics

Laser light is different from ordinary light. Ordinary light is **incoherent;** that is, its phase is inconstant, and part of the result of canceling itself out is that it appears diffuse. Laser light, on the other hand, is **coherent,** with a constant phase. Because of this, a laser beam leaving the laser through a pinpoint aperture will arrive at a wall opposite the laser and still have the same pinpoint size. This constant phase provides a greater degree of concentrated energy. It is important to make sure that lasers are never flashed in anyone's eyes, because laser light can cause eye damage quickly.

TABLE: PROPERTIES OF LIGHT

Ordinary Light	Laser Light
Small amount of energy	Huge amount of energy
Many different wavelengths	Single wavelength
Scatters out	Does not scatter out (except over interplanetary distances, and then slightly)
Many colors mixed	Single color
Cool	Sometimes hot; can melt metals with high-energy beam
Hard to focus in narrow beam	Narrow and extremely fine beam

Due to their unique properties, lasers are used in surgery, industry, consumer products, and space communications. Nowadays, grocery stores have laser scanners at checkout counters for reading bar codes. Music and video are recorded on disks that read music and video with laser light. The police use laser guns to check traffic speeds. Laser light is used to make 3-D (three-dimensional) photographs called holograms. Holograms are used on credit cards, postage stamps, and driver's licenses, and for many other applications.

(continued)

Materials: small helium/neon (HeNe) laser or laser pointer, large flashlight

- Shine the large flashlight on the ceiling of your darkened classroom, and move the beam away from you. As it moves, it will spread out considerably. This illustrates regular (incoherent) light. Shine the laser light on the ceiling, and move it away from you. The laser light is coherent. Students will easily be able to see the difference between the two.

Special Safety Consideration: Lasers can cause eye damage if students look directly into them. For this reason, keep lasers away from students when not in use. Both HeNe lasers and laser pointers are otherwise harmless—it will not burn or injure a student to have laser light fall on his or her skin.

Lenses transmit light. In the process, they focus light rays. Some lenses transmit light as is, while others make objects appear either enlarged or reduced. Microscopes are used to enlarge the view of small objects.

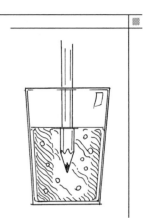

Materials: tumbler, some water, pencil or key, paper clip

- Build a simple microscope (one lens) by filling a glass with water and dipping an object, such as a pencil or a key, into it. Note how the object is enlarged. (Bottom lighting will improve the effect.)

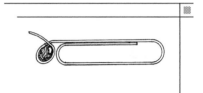

- Give each student or group a paper clip. Have students bend each clip to form a small loop on one end. Have them fill the loop with a small drop of water. They can use it as a magnifier to look at their hands or anything around them.

When light goes from one medium (transparent substance) into another, refraction occurs. Refraction is the deflection from a straight path undergone by a light ray or energy wave in passing from one medium (for example, air) into another (for example, glass) in which its velocity is different.

Materials: tumbler, pencil, water

- Fill a glass nearly full of water and place a pencil in the water at an angle. Notice how the pencil seems to bend. This apparent bending is refraction.

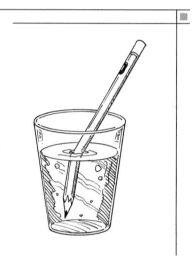

Materials: beaker, water, coin

- Place a coin at the bottom of a beaker nearly full of water. Holding the beaker slightly to the side, look down at the coin. The coin will appear to be where it is not, and you will see two coins. Shake the glass slightly to see which coin does not move. Notice that the real coin appears larger, since the water acts as a lens and magnifies the image.

Like any wave, light has wave fronts that are generated by the light as long as the light source is illuminated. These wave fronts interfere with one another, sometimes constructively, producing bright areas, and sometimes destructively, producing dark areas. In 1801, science knew nothing of this phenomenon until Thomas Young set up an experiment in which he directed light through two slits and observed it for the first time. Young, in 1801, did not have a coherent light source to work with (he used sunlight), so his experiment was slightly more complicated than ours will be.

Materials: laser pointer, mount for laser pointer (could be a large ball of clay), card with two narrow slits placed less than 1 mm apart, mount for the card, large sheet of paper, pencils

Procedure:

1. Mount the laser and the card so that the beam shines directly through the double slit on the card.

2. On a wall or book near the apparatus, mount a sheet of paper so that the image from the laser falls directly on it.

3. Dim the room, and turn on the laser. You will see bright lines interspersed with dark lines. These are called "fringes." Bright fringes occur when waves from both slits arrive in phase; dark areas result from the overlapping of waves that are out of phase.

4. Carefully sketch them onto the paper.

5. Turn off the laser.

Conclusion: Explain, in your own words, what causes the fringe pattern you saw on the paper.

Strong Safety Warning: While this type of laser is harmless in general, it can cause serious eye damage if you look directly into its rays. Make sure that neither you nor the other students working in the vicinity are in the line of the lasers.

The light we see combines all the colors put forth by the light's source. To break the light into its individual component colors, you use a **prism,** a triangular piece of glass or plastic, or a **diffraction grating.** A diffraction grating is a piece of plastic or glass with thousands of fine lines scratched on its surface. Rainbows demonstrate this process in nature; drops of rainwater, acting like prisms, break the sunlight into its component colors.

Materials: two pieces of cardboard (at least one of them white), razor blade, masking tape, two prisms (or one prism and a magnifying glass), film projector, black paper, clay

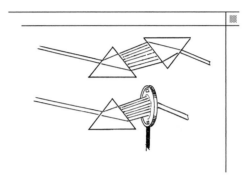

The following activities are done either with sunshine or a projector. If sunshine enters your classroom, do part A. If sunshine does not enter your classroom, then do part B. If you do not have a prism, do part C and use either light source.

- Perform the following procedure.

  1. Tape onto your window a piece of cardboard with a slit about 1 inch high and 1/8 inch wide. Place a table nearby so that the rays of the sun, going through the slit, hit the table. Make a base for the prism out of the clay; position the base and prism in the light on the table. Turn the clay until the prism projects a band of rainbow colors onto the piece of white cardboard, which you hold nearby.

  2. Place the magnifying glass or the second prism in the path of the rainbow colors and project them through to the white cardboard. This should recombine the colors together into the original light.

- Repeat all procedures as in part A, except use a projector as your light source. Cover the lens with the black paper and cut in it a vertical slit, as fine a one as you can make.

(continued)

Materials: water, small pan, mirror

- Fill the pan with water and place the mirror under the water, leaning on one side of the pan. Let the light come through the slit and strike the mirror. The mirror will project the spectrum on the wall below the slit. It may take a little effort to aim the mirror.

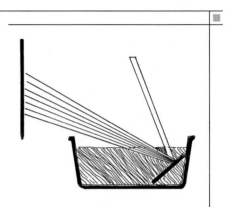

At a certain angle, light can no longer emerge from transparent substances. In glass, this angle, called the **critical angle,** is about 43°. If you shine a light into a triangular prism from the hypotenuse side, all the light will be reflected back at you, and none will pass through the glass to the other side. This phenomenon is called **total internal reflection.**

In this lab, you will determine the critical angle for water.

Materials: deep opaque tub, such as a galvanized washtub; waterproof flashlight; protractor

Procedure:

1. Fill the tub with water.

2. Put the waterproof flashlight under water, and turn it on with the beam shining straight out. Slowly turn it until the lightbeam does not exit the water. Try not to disturb the water too much, as this may give you a false reading.

3. Have a second student measure the angle of the flashlight to the bottom of the tub with the protractor. This is the critical angle. All the light is being reflected back into the tub, and this is known as total internal reflection.

Conclusion: What was the critical angle for water?

Potential energy simply means energy at rest, ready to become motion energy **(kinetic energy)** to do some work.

Materials: two books, marble

- Place a marble on a ramp made by a pair of books. While you hold the marble near the top of the book, the energy of the marble is potential. It has energy because of its position (height) above the bottom of the book. When you let go of 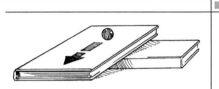 the marble, it rolls down and away because the potential energy has changed into motion energy—kinetic energy. Objects that are raised from the ground have potential energy. When they fall, they have kinetic energy. The fall is due to the pull of gravity. Thus, objects above the ground have potential gravitational energy.

Potential energy is increased if the height of the marble in the previous demonstration (26) is increased by making the ramp steeper. Height is the key variable. Work is done to bring the marble up the ramp. The higher the marble, the more work it took to get it there; thus, the more the stored energy it has.

Materials: empty plastic gallon bottle, pencil-type soldering iron, water, sink or suitable container to catch water

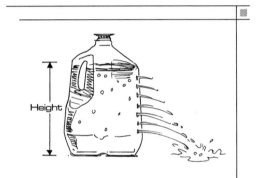

- Take an empty plastic gallon container and make about six to eight equally spaced holes from the bottom to the top. (A pencil soldering iron will melt circular holes.) Fill gallon with water rapidly, and observe how the water spurts out of the holes. As the height of the water column increases, so does the pressure, and thus the spurts get longer. The height of the water column creates the potential energy; the length of the spurts shows the kinetic energy.

Special Safety Consideration: Soldering irons can cause severe burns. Take care when using a soldering iron, and never allow students to use one in the classroom.

Speed is distance traveled per unit of time (e.g., per second). If the speed is in a specified direction, then it is called **velocity.** To find the average speed, divide the distance traveled by the time taken to travel it.

Materials: Frisbee™, measuring tape or meterstick, calculator, marble, several books, masking tape, stopwatch

- Throw a Frisbee from a reference line and measure how long it takes to land. Measure its distance from the starting point. Calculate the speed by dividing the distance by the time in seconds.

- Make a ramp on a flat surface with a pair of books, and mark distances in 1 m increments from it. Launch a marble down the ramp and time it as it rolls. Divide the distance by the time to obtain the speed. For ease of calculations, measure the time from 0 to 1 m, 1 to 2 m, 3 to 4 m or any other choice. In this experiment, we are looking for the average velocity; however, the marble is not progressing at a constant velocity. It is accelerating.

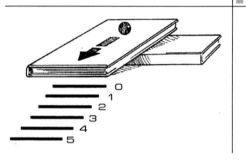

- To demonstrate **friction,** or resistance to motion, redo the same activity on a rougher surface, like carpeting or concrete, and compare the results. Friction will slow down everything.

Any object moving with a particular speed in a particular direction is said to possess velocity (*v*). Speed is a **scalar** number—that is, it reflects a certain number of km/hr, or cm/sec, or some other distance/time relationship. It cannot be a negative number. Indeed, we can only speak of speed as an absolute value. Velocity, on the other hand, has a particular direction, and therefore could be considered a negative number. We call this type of quantity a **vector.** Velocity also has a particular beginning and end point, which speed does not have. Velocity can be defined as a change in distance over a change in time.

In this experiment, you will run three trials. In the first case, you will see and calculate a simple constant velocity, starting from zero and proceeding forward. In the second, you will see a constant velocity starting somewhere other than zero and proceeding forward. In the third, you will measure a constant velocity starting at 600 cm and proceeding backward.

Materials: small battery-driven toy car, six meter sticks, stopwatch, small bits of paper, graph paper, pencils

Procedure:

1. Lay the meter sticks down end to end, so that each "100" is followed by a "1."

2. With your partners, let the car begin at the first meter stick at zero cm and continue forward. Have a student call out "time" every two seconds, while a second student drops a small ball of paper at the place where the car was when the first student called out.

3. When the car has finished its trip, create a graph. On the *x*-axis, write down the times in seconds, so intervals read "0, 2, 4, 6 . . ." On the *y*-axis, record intervals of 100 cm, starting at zero. Then plot on the graph the positions of the car at each time interval. Connect the points and draw the graph. What is the shape of the graph? Where does your graph begin? Where does it end?

(continued)

4. Now, take the distance between two time intervals, and divide by 2. For instance, if between 2 seconds and 4 seconds the car traveled 90 cm, determine its velocity by dividing 90 cm by 2 sec.

5. Repeat the exercise, but start the car at 40 cm. How does this change your findings and your graph?

6. Repeat the exercise, but start the car at 600 cm and have it drive in the opposite direction. How does this affect your graph?

In the first case, you have a positive velocity, as represented by the positive slope. In the second case, you also have a positive velocity, but the y intercept is not at zero, as it was in the first case, but at 40. In the last case, you have a negative velocity as represented by the negative slope. In each case, however, the speed (cm/sec) you calculated should be pretty similar.

Conclusion: How do speed and velocity differ? Why do we differentiate between them?

Acceleration is the change in speed per unit of time. (Note that acceleration can refer to something slowing down; this is known technically as *negative acceleration*, but is generally called *deceleration*.) To calculate acceleration, subtract the original velocity from the final velocity and divide the result by the time. One unit of velocity is m/sec. One unit for acceleration is m/sec/sec or m/sec^2. The second power is your clue that the unit is acceleration. Acceleration is a gauge of how rapidly speed changes. Racing cars can go from 0 to 60 mph in a few seconds, while most cars take longer. The difference is the racers' greater ability to accelerate, due to their more powerful engines.

Materials: books, marbles, masking tape, metric ruler, stopwatch

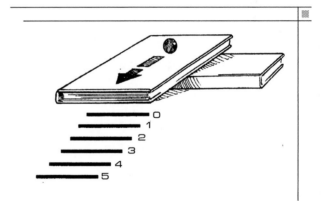

- This demonstration will take a full period. Set up a ramp system like the one shown in the illustration. Make certain that you have at least 5 m of run. Test the ramp to make sure your marble will roll at least 5 m. I use a 5 m counter. The floor is fine if it is smooth. Otherwise try a spot outside your classroom. The entire class participates with you in this demonstration. Some students are the timekeepers, some are launchers, and some collect the marbles, while everyone else records the data and calculates the results. The goal is to find the average speed for each 1 m interval. Repeat all measurements three times and then average the three. The difference in the average speeds between each interval is the average acceleration for that interval.

(continued)

Copy and complete the following data:

	Time	Time	Time	Time	Time
	0–1 m	1–2 m	2–3 m	3–4 m	4–5 m
Launch 1					
Launch 2					
Launch 3					
Average Time					
Speed	A	B	C	D	E
Acceleration	A–0	A–B	B–C	C–D	D–E

The capital letters in the boxes represent the calculated speeds. To find the accelerations, subtract the speeds as indicated. The final unit will be m/sec/sec or m/sec^2. Note that the original speed is 0.

Galileo developed the concept of acceleration in his experiments on inclined planes. His main interest was free-falling objects, but, because he lacked suitable timing devices, he used inclined planes to slow objects down. The ramps helped him investigate accelerated motion more carefully. Galileo found that a ball rolling down an inclined plane will pick up the same amount of speed in successive seconds. That is, the ball will roll with *unchanging acceleration*. For example, a ball rolling down a plane inclined by a certain angle might be found to pick up a speed of 2 meters per second for each second it rolls, or a constant acceleration of 2 m/sec/sec, or 2 m/sec^2. Its speed at 1 second intervals at this acceleration would be 0, 2, 4, 6, 8, 10 m/sec, and so forth. Therefore, you can see that the speed of the ball at any given time after being released is: Velocity = acceleration × time, or $v = at$.

If we substitute the acceleration of the ball in this relationship, we can see that at the end of 1 second, the ball is traveling 2 m/sec, and at the end of 2 seconds, it is traveling 4 m/sec.

If we know our final, or *terminal velocity*, we can solve for acceleration instead. First, we have to find our terminal velocity, a simple enough matter. We divide the length of our inclined plane by the time it took for our sphere to reach the bottom. Then, by plugging velocity and time into their variables, we can solve for acceleration. Divide our terminal velocity by time, and we will arrive at the acceleration rate in meters per second per second (m/sec^2).

Materials: long, smooth board (measure in centimeters), books or other objects to create an inclined plane, stopwatch, protractor, spheres of various sizes and masses

Procedure:

1. First, as a visual demonstration of the law of falling bodies, release two or more spheres at the top of the inclined plane. They will arrive at the bottom of the inclined plane together, regardless of their mass.

(continued)

2. Accurately measure the angles of the inclined plane. Perform the experiment on three different angles—30°, 45°, 60°.

3. Select one sphere to be used throughout the experiment.

4. Do three trials at each angle, and create graphs for speed, based on rate of acceleration, at 1 sec, 2 sec, 3 sec (may have to be projected).

5. Create a graph of the acceleration versus the angle.

6. What pattern do you notice? You should see a higher acceleration rate for steeper angles.

7. Can you determine what the acceleration rate will be if the angle is 90°? (9.8 m/sec^2)

Conclusion: What does the law of falling bodies imply about inertia—an object's resistance to changes in velocity? Can you predict at what angle an object with no motion initially would begin to accelerate? Why?

Momentum is described as mass times velocity. The physics shorthand for momentum is $p = mv$.

Some collisions are elastic. Consider billiard balls on a pool table. If a billiard ball hits another head on, then the first ball will stop and transfer all its momentum to the second, which will roll away at the same velocity the first ball was moving at when it struck the second. *Others are inelastic*—the two objects become entangled after the collision. Consider the case of two freight cars on a railroad track. If they are the same mass, and one is at rest, while the other is traveling at 10 km/hr, after the inelastic collision, the two will continue moving, albeit at a slower 5 km/hr.

In all collisions, momentum is conserved. The momentum may now be spread among several objects, transferred from one object to another, or may even become heat and deformation, but all momentum can be accounted for physically.

Materials: long piece of Hotwheels™ track, as well as the launcher that came with the set; two vehicles, one more massive than the other, that fit into the track without serious problems with friction; meterstick; stopwatch; balance scale; tape

Procedure:

1. Tape the meterstick along the track.

2. Measure the mass of the two vehicles.

3. Put the more massive vehicle in the launcher and send it down the track.

4. Calculate its velocity and its momentum.

5. Put it back in the launcher, and place the less massive vehicle on the track not too far from the launcher. Be sure to record its position.

(continued)

6. Send the vehicle down the track again and allow it to collide with the less massive vehicle. Observe what takes place.

7. Now, put one piece of tape on the front of the massive vehicle and one on the back of the less massive one. Put the two cars back into their starting positions, and release the massive vehicle again.

Conclusion: What was the difference between the first trial and the second? Another way to try this is to have the smaller vehicle be the striking vehicle. What differences would you expect to find?

Bernoulli's principle states that when the speed of a moving fluid *increases*, the pressure on its edges *decreases*. This principle explains why planes, birds, and kites fly. It explains many things we see in our lives, from the action of rocket motors to the action of paint spray guns.

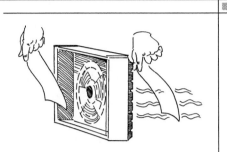

Materials: electric fan, strip of paper, Ping-Pong™ ball, funnel, small box full of polystyrene packaging particles (peanuts), vacuum cleaner hose 5 to 8 feet long, hard-boiled egg, glass, deck of playing cards

- First, demonstrate the effects of high and low air pressure: Turn on the fan and show the paper strip in front and behind the fan blades.

- Put the Ping-Pong ball in the top of the funnel, and blow upward into the stem of the funnel. The ball will not blow away but will cling to the funnel. Invert the funnel so that it points downward. Hold the ball temporarily. As you start blowing hard, slowly remove your hand. Again, the ball is going to cling to the funnel. As the air moves away from you faster, it creates a low-pressure area in the center.

- Take a piece of vacuum hose and place it in a box containing polystyrene particles. Spin the opposite end of the hose in a circular motion. As you increase the speed of the end, the pressure will drop and pieces of foam will come out as though sucked up by a vacuum cleaner. This is an outstanding demonstration, but it requires a cleanup.

- Place a hard-boiled egg in a glass and move the glass under a faucet open at full power. (Remove the aerator if there is one; this activity needs a solid stream of water.) Maneuver the egg into the

(continued)

water stream. The egg will rise against the stream of water. The increased speed of water splitting around the egg causes lower pressure.

• Ask one of your students to hold a playing card and try to make it fall straight down. The student will have trouble with this. Now, hold a playing card horizontally. Let it go. It will come practically straight down. Hold the card on its side. It will come down flipping over and off to the side. A flat card is a wing. As it falls, it compresses air below it, and above it there is less pressure. A vertical card creates a turbulent flow and does not fly.

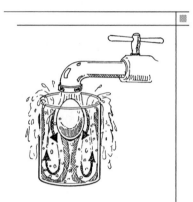

Newton's third law of motion states that for every action there is an equal and opposite reaction.

Materials: two balloons, string, soda straw, masking tape

- Inflate a balloon. Hold it pinched so that no air escapes. Inside the balloon, air exerts the same pressure on all sides of the balloon. Let go of the balloon and observe how it flies around erratically until all the air inside has gone out. As the compressed air (a region of higher pressure) escapes into a region of lower pressure, it creates a force to provide balance (equilibrium). An equal and opposite force pushes the balloon forward.

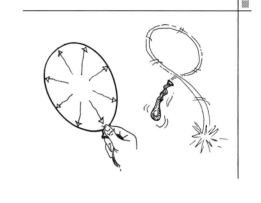

- Stretch a piece of string across the classroom. Before tying the string taut, insert the string into two 2-inch soda-straw pieces. The string should be roughly horizontal, but pitching it slightly downward, in the direction of travel, will help. With the assistance of students, tape an inflated (pinched but not tied-off) balloon to the two pieces of soda straw with the balloon's exhaust parallel to the string. Be careful not to attach the masking tape to the string. Pull the entire assembly to the end of the string and let go. The balloon should be able to cross the room. (It may take some practice.)

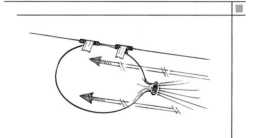

Magnets attract only iron, cobalt, nickel, and some materials made with these elements. A bar magnet's end is a **pole.** One is the north pole and the other is the south pole. Magnetic poles follow the same rules as electric charges: Like poles repel, unlike poles attract.

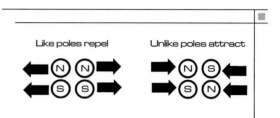

Materials: two bar magnets, paper clips, string, pencil, masking tape

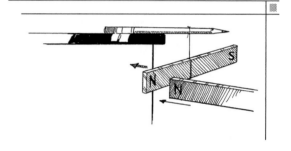

- Tape the pencil to the edge of your table. Tie a magnet to the pencil so that it balances. It will slowly align itself along the north-south meridian and act as a compass. Take the other magnet and bring one end of it closer first to the north pole and then to the south pole, showing how it attracts one and repels the other pole.

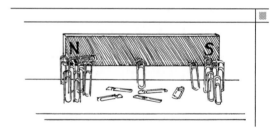

- Put a handful of paper clips along a line. Place the magnet over them and lift the paper clips. More paper clips are picked up at the ends, showing that there is more magnetic force near the poles.

Atoms are like tiny magnets turned in random directions. If they are all lined up, their magnetic forces combine to make a material (iron, nickel, or cobalt) magnetic. To magnetize an object made of iron, nickel, or cobalt, gently pull a magnet along its length, and do not remove the magnet until it reaches the end. Repeat the stroke at least 10 times. You are lining up the atoms inside the object. By stroking it several times, you are increasing the number of atoms lined up. When atoms are lined up, their magnetic forces add up. When you stroke a nail from head to point with the north pole of a magnet, the point of the nail will become a south pole. Most magnets that you prepare from iron materials will not keep their magnetism very long. Magnets made from *alnico* are much stronger magnets, and they keep their magnetism for years. Alnico is made of nickel, aluminum, cobalt, copper, and iron.

Materials: bar magnet, several nails, paper clips

- Use a nail and show that it does not lift a paper clip. Stroke the nail with the magnet 10 times. Lift one or two paper clips with the nail.

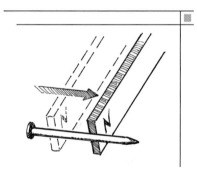

When a magnet touches a material that transmits the magnetic force, it creates magnetic induction. Iron and steel are such materials.

Materials: bar magnet, several paper clips

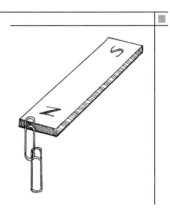

- Hang one paper clip from the magnet. Gently pick up another paper clip by using the end of the first paper clip. The magnetic attraction is stronger than the force of gravity, and the paper clips will stick to each other.

Magnetic **force lines** are invisible, and they travel from one pole to the other without touching each other. All the magnetic force lines together are a **magnetic field.**

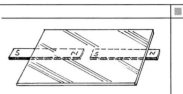

Materials: two bar magnets, horseshoe magnet, iron filings, a piece of stiff acetate or a sheet of glass, overhead projector

- Place the bar magnets on your overhead projector with the opposite poles about 1 inch apart. Cover them with a sheet of acetate. Sprinkle the sheet with iron filings from at least 1 foot up. Gently tap the acetate, and the lines of magnetic forces will become visible. Refocus your projector to show the magnetic lines and the magnetic field. Repeat with like poles, either N-N or S-S, or with a horseshoe magnet.

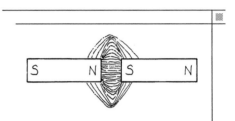

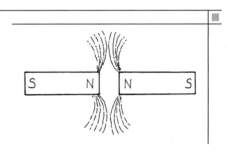

Our Earth is a giant magnet. Earth has two large magnetic poles, one near the South Pole, the other in Hudson Bay about 1,000 miles away from the North Pole. Earth's magnetic field stretches far out into space. It protects living things on the planet's surface by deflecting particles streaming from the sun. Without a magnetic field, there would be no life on Earth. The planet has magnetic lines that affect all magnetic materials. Deep in the abyssal depths of the Atlantic Ocean, ferrous particles, frozen in former magma near the edges of tectonic plates, clearly indicate from their orientation Earth's magnetic lines and their reversals over millions of years.

It is possible to make a compass to detect Earth's magnetic field.

Materials: pencil, string, tape, bar magnet or horseshoe magnet, masking tape, sewing needle, cork, small bowl, water

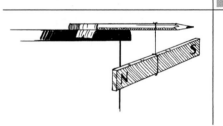

- Tape a pencil to the edge of a table, and tie a bar magnet to the pencil so that it balances. (Alternatively, use a ruler stuck between the pages of a book. Cover the book with a couple of additional books to hold things in place.) The bar magnet will line itself up along the line of Earth's magnetic field.

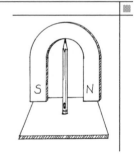

- If you wish to use a horseshoe magnet, take a small piece of wood, about 3 inches square, and drill a hole in its middle to fit a pencil. Glue the pencil in place. Balance the horseshoe magnet on the pencil tip. Slowly the magnet will line itself up along Earth's magnetic meridian.

- Stroke the sewing needle a dozen times until it becomes a magnet. Push it through the cork and place it in a small bowl nearly full of water. You have just made a water compass.

To destroy the magnetism in a magnet, the magnet must be hit, dropped, or heated. Hitting and dropping a magnet causes catastrophic disarray in the molecular arrangement, whereas heating one causes molecules to move away from one another, destroying the magnetic force. Magnets are dipoles, so breaking a magnet into two only makes smaller magnets. Magnetic tapes are demagnetized (erased) by bulk demagnetizers. These magnetize the tapes at 90° to the original lines, so that the magnetic reading heads in VCRs and tape recorders do not read the magnetism.

Materials: coat-hanger wire, strong magnet, compass, hammer, wire cutters, tongs, Bunsen burner

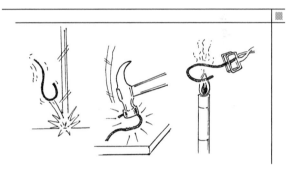

- Cut a piece of coat-hanger wire. Check with the compass to see that it is not a magnet. Magnetize it by stroking it with the magnet a dozen times. Check with the compass to see that it has become a magnet. Drop the wire several times until it becomes demagnetized.

- Remagnetize the wire, performing the compass tests as above. Hit the wire several times with the hammer. Again, it has lost its magnetic force.

- Remagnetize the wire and test it. Hold it in the flame of the Bunsen burner with the tongs. Again, it becomes demagnetized.

Static electricity is electricity that is collected by rubbing certain surfaces together. *Electricity* is the flow of electrons. Electrons like to move about freely. As two surfaces rub together, electrons leave one surface and collect on the other. The surface that has the excess electrons is *charged*. If a charged surface makes contact with the ground, then the excess electrons discharge, and sparks often result. Humidity acts as a conductor and discharges static electricity most of the time. The demonstrations below work best when it is very dry and cold or hot. Students can relate to going across a carpet and getting a mild shock as they touch a doorknob. To avoid this uncomfortable experience, touch the doorknob with a key first. The same spark will happen, but the key feels no pain.

Materials: two balloons, string, piece of fur (or silk or nylon), small fluorescent tube, paper

- Inflate a balloon and tie it shut. Darken the room and allow time for everyone's eyes to become adjusted. Rub the balloon with the fur and touch the fluorescent bulb. Sparks will seem to move from the balloon to the bulb, then the bulb will begin to glow faintly.

- Cut a small piece of paper into very small pieces, as small as any letter in this text. Rub the same balloon with the fur and touch the paper. The paper will stick to the balloon.

- Inflate the second balloon. Hang both balloons together. Charge them both. The balloons will hang away from each other, because like charges repel. Both balloons are charged with electrons, which are all negatively charged. Place a student between the two balloons. The balloons will come together, as though they were kissing the student. Their charges discharge through the student.

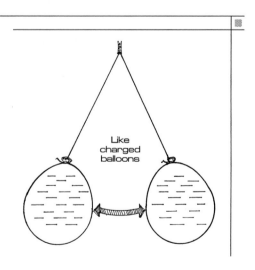

Like charged balloons

The Van de Graaff generator is a machine that generates static electricity, making it very useful in demonstrations. It consists of a small motor that turns a belt at about 3,600 rpm. The rapidly moving belt touches a brush, collecting static charges, and transfers the charges to a spherical dome. In theory, the machine can generate about 5.3×10^{-9} amperes; but in practice, it makes only about 2 to 2.5 microamperes—far below the safety limit. The generator provides about 200,000 volts when its belt is 1 inch wide. The rubber belts will operate for about 20 hours, and they work well in humidity above 75%.

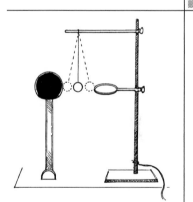

Materials: Van de Graaff generator, fluorescent tube (small), Ping-Pong ball, stand with ring, string, adhesive tape, several small strips of silk or other conductor, copper wire, polystyrene block, pointed metal object

- Attach the Ping-Pong ball to a string with adhesive tape. Hang it from a stand clamp. Install the ring at the same height as the dome of the generator. Position the hanging ball between the ring and the generator. Wrap one end of a piece of copper wire with stripped ends to the stand, and either tie the other end of it around a water pipe or stick it into the ground hole in an electric outlet. (**Caution: Do not connect the wire unless you are 100% sure that you have the right hole.**) Gradually move the generator closer to the ball until the ball moves; it will be attracted and then repulsed and pushed to the ring. The ball will scurry back and forth.

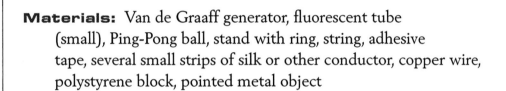

- Darken the room and bring the fluorescent bulb close to the working generator. It will begin to glow.

(continued)

- Hang a few pieces of silk above the generator and slowly bring them closer, until they all charge and start being repelled in all directions.

- To raise a person's hair, have a volunteer stand on a large block of polystyrene insulation and touch the generator with a pointed metal object. (Direct contact with fingers may be uncomfortable.) After a few seconds, you will observe the hair standing up a few strands at a time. For the most striking results, choose a person with freshly washed, light-textured hair, without any mousse or other chemicals.

Easy Science Demos & Labs:
Physics

A wire conducting electric current has a magnetic field around it. Hans Christian Oersted, a Danish scientist, discovered this phenomenon more than 200 years ago.

Materials: battery, transparent compass, coil of bell wire, stand, ring holder, two clamp holders, loose bell wire, small piece of cardboard about 4 inches square, four or more small compasses, iron filings

- Place a coil of bell wire on the overhead projector. Place the transparent compass in the middle of the wire. Connect one end of the battery to one end of the coil of wire. Touch the other end of the wire to the battery so that current flows through the wire. As you touch the battery, the compass deflects sharply, indicating the presence of a magnetic field. Do not leave the battery connected for too long, because you have a short circuit.

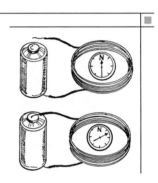

- Repeat the activity, placing a single wire across the compass and touching the battery. You will obtain the same effect as with the coil, but with less deflection of the compass. (This is Oersted's original experiment. He noticed the deflection, while others failed to observe it.)

- Punch a small hole in the center of the cardboard and pass the wire through it. The wire must be perpendicular to the cardboard. Place the cardboard on the ring. Clamp the wire at the top and bottom. Place the compasses around the wire, about 2 inches away. Connect the battery to the wire and observe how the compasses form almost a ring around the wire, indicating magnetic forces. Remove the compasses and sprinkle iron filings around the wire. With the wire connected to the battery, tap gently on the cardboard. You should obtain concentric rings.

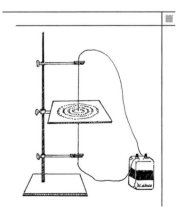

An **electromagnet** is a magnet only as long as electric current goes through the circuit. The magnetism from a short piece of wire is weak, so by winding many turns and changing the wire into a coil, one gets more magnetism. Electromagnets have poles like regular magnets. Electromagnets can be turned on and off, unlike regular magnets. Electromagnets are used to pick up junk metal, in making relays, and in solenoids. They control the water valves in dishwashers and the horns in cars, they open and close garage doors, and they have thousands of other applications.

Materials: two or three batteries, bell wire, iron filings, large piece of cardboard, two compasses

- Remove the insulation from the wire. Wrap the wire tightly around a pencil to form a long, tight coil. Sprinkle iron filings on the cardboard. Place the wire coil on the cardboard, in the middle of 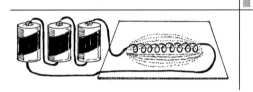 the iron filings. Connect one end of the wire to the negative end of the batteries in series, and connect the other end to the positive. Keep the electricity going until you get a pattern.

- Place a compass near each end of the coil and briefly connect the electricity to the coil. This will establish the coil's polarity. At the south pole, the compass needle points toward the coil. At the north pole, the compass needle points away.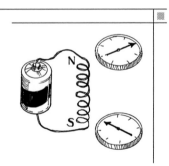

With a coil of about 20 turns, you may be able to lift one paper clip. To increase the magnetic strength of the coil, you can:

1. increase the number of turns.

2. place a piece of iron in the center of the coil, as its core.

3. increase the current in the wire.

4. any combination of these steps.

Materials: bell wire, long nail or iron rivet, batteries, paper clips

- Wrap about 20 turns of wire around the nail. Pull the nail out and turn on the electro-magnet. Try to lift paper clips with the coil. If you can lift one, you are doing well.

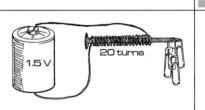

- Slip the nail back into the 20 turns of wire and turn on the electromagnet. Lift paper clips with the end of the nail. You will lift several.

- Wrap 50 turns of wire around the nail (you can start a second layer). Turn on the electromagnet and lift paper clips. You will be able to lift many more paper clips than before.

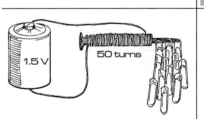

- Connect a second battery in series with the first. Turn on the electromagnet and lift paper clips. You will lift even more paper clips.

Magnetic force is generated around a wire when electric current goes through it. Conversely, if a magnetic field cuts across a wire, an electric current is induced in the wire. This principle is used to generate electricity and to make all types of electrical-generating equipment. The amount of electric current increases with:

1. the number of wire turns.

2. an increase in the magnetic field strength.

3. the speed of movement of the magnetic field.

Materials: projection galvanometer (if available) or a regular one, coil of bell wire, bar magnet, larger horseshoe or more powerful magnet, hand-crank generator, overhead projector

- Place the galvanometer on the overhead projector. Connect it to both ends of the bell wire loop. Holding the loop vertically, move the bar magnet forward and backward through the loop. Observe the increased reading as you move through the loop. Notice also that as you withdraw the magnet, the current appears to move in the opposite direction on the galvanometer. You have just made *alternating current*, or AC. Demonstrate that you can also hold the magnet still and move the coil of wire over the magnet with identical results.

Galvanometer

Magnet
Coil of wire

- Repeat this demonstration, moving the magnet in and out more rapidly. Notice the increase in current.

- Repeat the previous activity using the more powerful magnet. This time lay the loop flat on the projector and move the magnet up and down. Notice the increase in the electric current.

- If you have a hand-crank generator, have all students form a large ring, holding their hands together. At this point give

(continued)

them a mild shock. Explain to them that the generator is nothing more than a moving coil inside the magnetic field of four to six horseshoe magnets. First show how they receive no shock if they are in parallel with the bulb. Next open the circuit, put the students in series, and watch out!

Special Safety Consideration: Students with pacemakers should NOT participate. Inform students ahead of time that they will feel a mild shock, and give them the opportunity to opt out of this demonstration.

Working electromagnets can induce electric current in another circuit. If two coils share the same core and the primary coil conducts electricity, it induces magnetism in its core. When the shared core is magnetized, it produces electricity in the secondary coil. This is true for alternating current. With direct current, this happens only when you open and close the circuit. Electromagnetic induction is used in cars to provide the high voltage to fire the spark plugs. Electromagnetic induction is also used to step electric voltage up or down with transformers.

Materials: switch, battery, stripped wire, nail, galvanometer, battery holder

- Wrap 30 turns of wire around the upper half of the nail and leave about 6 inches of wire at each end. Do the same on the bottom half of the nail. Connect the upper wire to the galvanometer. Connect the lower wire to the positive pole of the battery and to the switch. Then wire the switch to the negative pole. Close the switch and observe the reading on the galvanometer.

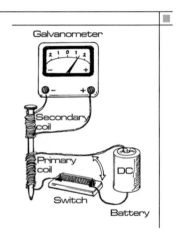

You will be assembling a working simple circuit with a switch. A lightbulb will only go on when a circuit is closed—that is, when the electricity has a conductor that can go all the way around in a circle.

Materials: 1.5-volt lightbulbs with attached leads (To save time, the teacher should bare the ends of the leads in advance.); D batteries; tape (Duct tape works best.); brass paper fasteners; scissors; metal paper clips; sheets of cardboard, about 15 cm × 15 cm

Procedure:

1. Take your lightbulb assembly and carefully tape one side all the way around a sheet of cardboard, so that the bulb is at the top and the exposed wires are at the bottom. Tape the exposed wire to the top of the D battery, and tape the battery securely to the cardboard.

2. Carefully cut the other lead in two, and very carefully scrape the plastic away from each end using the scissors, so that all ends have exposed wires.

3. Tape the lead coming from the lightbulb down the other side of the cardboard, leaving only about 2 cm free.

4. Punch a hole in the cardboard with one of the brass paper fasteners, and wrap the wires around the stem. Push the fastener through the hole and use the prongs to hold it to the cardboard.

5. Take the cut piece of wire and fasten one end to another brass fastener. Place the fastener as far as you can from the other and still have a paper clip touch both.

6. Tape the cut piece the rest of the way to the battery, and touch the exposed end to the bottom of the D battery. Tape it firmly on. The lightbulb should be off. If it is not, check to make sure that the paper fastener prongs are not touching under the cardboard. If they are, move them.

(continued)

7. Take the metal paper clip and lay it across both paper fasteners. The bulb should light—you have completed a circuit. Remove one of the brass fasteners and slip the paper clip through the prongs. Put the fastener back on. Now you have a working circuit with a switch.

Conclusion: What are necessary components to complete a circuit? Diagram your circuit.

A *circuit* is a path where electric current can flow. Circuits are opened and closed by switches. Switches can be made in many ways; there are knife-blade, push-button, sliding, toggle, mercury-type, and many other kinds of switches. The simplest switch is a break in a wire; you operate it by touching the wires together or pulling them apart. Electricity can flow in a circuit only if electrons leaving the negative end of the power supply can return to the positive end. The power supply can be either direct current (DC) or alternating current (AC). The purpose of a circuit is to operate something run by electric current, such as an appliance, a motor, a light, or a TV. **Series circuit** means that electric current has only one path and must go through all the components of the circuit.

Materials: battery, bulb with base, switch, wire to connect these components

- Connect the battery to the bulb, then to the switch and back to the battery. Show how electricity has only one path to follow. This is a series circuit. The bulb lights up only when the switch is closed. If you had several bulbs in series and one burned out, the circuit would be open, and none of the bulbs would work. The bulb would have acted as a switch. In series circuits, if any component fails, the circuit opens and does not work.

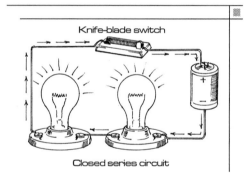

Closed series circuit

- Connect two bulbs in series, and close the switch. Both bulbs will light up. Gently unscrew one bulb. As the bulb opens the circuit, the other bulb will also go out.

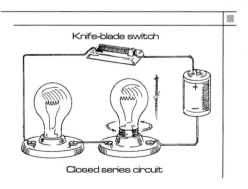

Closed series circuit

A **parallel circuit** is an electrical circuit in which electrons leaving a power supply have two or more paths they can follow to go back to the power supply. If several bulbs in a lamp were wired in parallel, and one failed, the others would continue to work. Your classroom lights are wired in parallel.

Materials: two bulbs with bases, battery (with holder, if needed), switch, enough wire to connect them

- Connect one end of the switch to the battery's positive pole, and connect the other end of the switch to both bulb bases. Connect the battery's negative pole to both bulb bases. Close the switch, and both light-bulbs will light up. Gently unscrew one bulb. The other bulb will remain lit and will shine a bit more brightly because the voltage is not divided between two bulbs.

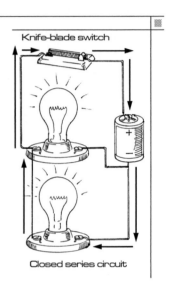

Knife-blade switch

Closed series circuit

A **short circuit** occurs when electric current from a power supply can return to the power supply without going through any resistance. A common example: Two wires of a plugged appliance touch. Short circuits are dangerous, for they start fires. To protect people, all electrical circuits have fuses, which will open the circuit if a short occurs. A new type of fuse, a ground-fault circuit interrupter, has been developed. This fuse detects minute leaks of current to ground, and by opening the circuit before a large current electrocutes a person, it saves lives. The conventional fuse trips only with large currents, by which time a person may have been electrocuted. Most appliances are grounded to provide a safe return to ground of leaking electrical currents. Modern electric plugs have polarized ends, with one prong wider, to provide electrical safety in case of appliance failure.

Materials: battery, bell wire with insulation

Short circuit

- Connect both bare ends of an insulated wire to a battery. Tap one end on and off and do not let it stay connected. Otherwise, in a short time you will notice that the wire is getting hot and the battery is also getting hot. Since there is no resistance in the wire, an increasingly large current tries to go through, heating the wire. The battery will heat up in trying to supply infinite current. In the process, the battery will eventually go dead.

A *volt* is a measure of electrical pressure and is named after the Italian scientist Alessandro Volta, who did much basic work with electricity. To measure **voltage,** you need a voltmeter. Electrical pressure is another term for electromotive force. A voltmeter must always be connected in parallel with any circuit it measures. A voltmeter can be digital (DVM) or analog (with a moving needle). Note: You can also measure the voltage of a battery, but the result is inaccurate. To get a proper reading, place the battery in its circuit, and while the battery works, measure the voltage. A value of 10% below the battery's rating is fine. An alternative is to have a meter specifically designed to measure batteries.

Another instrument increasingly used in high schools and colleges for measuring resistance, voltage, and current is the digital multimeter (DMM). DMMs are not superior to basic analog meters or individual meters for measuring the three quantities, but they are becoming the industry standard. DMMs are the instruments students are most likely to see in engineering programs, physics departments, and technical programs. Should you be in the enviable position of having to make purchases for your science department, consider the handheld digital multimeter for your students.

Materials: battery, voltmeter or multimeter, bulb, switch, small electric motor (optional)

- Read the instructions provided with the meter. Prepare a circuit, and then close it and measure the voltage across the battery.

- Measure the voltage across the battery out of circuit.

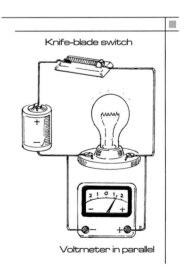

Knife-blade switch

Voltmeter in parallel

Electric **current,** the quantity of electricity going by a point in a circuit, is measured with an ammeter. The unit is the *ampere*, named after the French physicist A. M. Ampere, who did much basic work with electricity. To measure the current in a circuit, the ammeter must be in series.

Materials: bulb with its base, wires for connections, battery, several resistors, ammeter or multimeter, switch

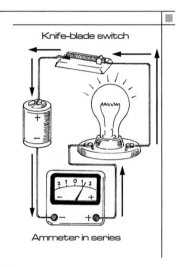

Knife-blade switch

Ammeter in series

- Assemble the circuit and place the meter in series. Measure the current used by the bulb and the resistors.

The **resistance** to the passage of electric current is measured with the ohmmeter and the unit is the *ohm*, named after the German scientist Georg Ohm, who did much basic work in electricity. **Insulators** oppose the flow of electric current and have a high resistance. **Conductors** have a low resistance. Factors that affect the resistance of wires are:

1. the thickness of the wire.

2. the length of the wire.

3. the material of the wire.

4. the temperature of the wire.

Wires are made from several metals. Aluminum and copper are good conductors because they have a low resistance. Toasters and electric heaters are made with wires of *nichrome,* a metal that has a higher resistance than copper and aluminum and gets hot when electric current goes through it. *Tungsten* also has a high electrical resistance and is used as the filament for lightbulbs. It gets so hot from electric current that it glows. Metals improve their conductivity if they are cooled. Materials that lose their resistance at low temperatures are **superconductors.** Mercury, a good conductor at regular temperatures, becomes a superconductor at −454°F. Ohmmeters have internal batteries that provide power for the resistance test. Never test any circuit with power in it. Never test a battery or a live outlet for resistance.

Materials: ohmmeter or multimeter, several resistors

- Measure the resistance of several resistors and compare your measurement to the posted value of the resistors.

Georg Ohm figured out a basic law for electricity: $V = I \times R$ where:

> V = Volts (electric potential)

> I = Amperes (current)

> R = Ohms (resistance)

If the formula is solved for I, then $I = \frac{V}{R}$. This relationship provides much information. Current is proportional to voltage; therefore, increasing the voltage increases the current. The current is inversely proportional to resistance; therefore, as resistance increases, current decreases. The graphic above is a convenient way to remember **Ohm's law:** Cover what you need and multiply or divide the other two values.

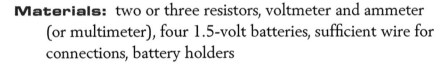

Materials: two or three resistors, voltmeter and ammeter (or multimeter), four 1.5-volt batteries, sufficient wire for connections, battery holders

- Wire the batteries through the resistor and switch. Measure the current by placing the ammeter or multimeter in series. Measure the voltage by placing the voltmeter or multimeter in parallel. Take the resistors out of circuit and measure their resistances.

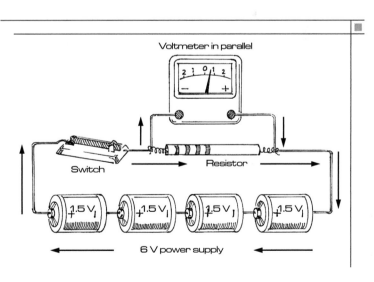

(continued)

- Try the same activity, using less than 6 volts. Try it with two or three batteries. Check that the measurements match the computations. Use Ohm's law to verify values.

Here is a table of typical values measured and calculated.

Volts	Ohms	Amperes
6	12	0.5
6	3	2
6	60	0.1
4.5	100	0.045
4.5	50	0.09
3	300	0.003
3	3	1

Temperature is a measure of the speed of molecules. Several different units are employed to measure it. A *thermometer* is a device that measures the temperature of matter. Celsius and Fahrenheit systems are used quite commonly in the United States. Celsius and Kelvin scales are recognized worldwide. A table of temperature conversions is found in the Appendix. To convert Celsius and Farenheit temperatures, the following formulas are used (F° is degrees Fahrenheit, C° is degrees Celsius):

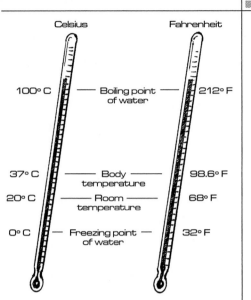

$$C° = \frac{(F° - 32)}{1.8} \text{ and } F° = (C° \times 1.8) + 32$$

There are 100 degrees in the Celsius scale and 180 in the Fahrenheit scale between the points at which water freezes and boils. From there we get the 1.8 factor in the formula. The ± 32 is the adjustment for the freezing point in the Fahrenheit system, because 32°F is equivalent to 0°C.

Materials: thermometer, tap water, hot water, cup

- Place some tap water in the cup and measure its temperature. Empty the cup and fill with hot water. Measure its temperature. In this demonstration, a camcorder with close-up image on a monitor is very helpful. Comment to your students that the expansion of the liquid in the thermometer is a property of heat. Solids, liquids, and gases expand when heated.

A **calorie** is a measure of how much heat must be added or taken away to make something heat up or cool down. One calorie raises the temperature of 1 g of water 1°C. It would take 15 calories to raise the temperature of 1 g of water 15°C, or 15 calories would raise 15 g of water by 1°C. One kilocalorie (1000 calories) equals 1 food calorie. This value provides a clear way of establishing the energy in foods. See Appendix for sample values.

Materials: three pencils or dowels, masking tape, empty soda can, Celsius thermometer, water, aluminum plate, clothespin, T-pin, peanuts, matches, graduated cylinder

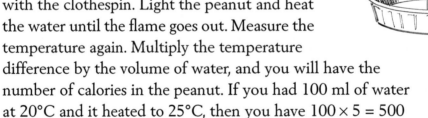

- You will measure the caloric content of a food by burning the food and using its energy to heat water. Raise the soda can on stilts, using the three pencils or dowels fastened with tape. Place 100 ml of water in the can. Stand the can in the plate. Place the thermometer in the can and measure the temperature of the water. Skewer a peanut with the T-pin and hold it with the clothespin. Light the peanut and heat the water until the flame goes out. Measure the temperature again. Multiply the temperature difference by the volume of water, and you will have the number of calories in the peanut. If you had 100 ml of water at 20°C and it heated to 25°C, then you have $100 \times 5 = 500$ calories or .5 food calories.

The second law of thermodynamics relates to heat transfer. Unless acted on from outside, heat always flows from a region with a higher temperature to a region with lower temperature. The system has to be "closed" to see this happen. For instance, if you were to leave a cake cooling on a hot stove, although the external air would be cooler, there would be an energy flow into the cake from the stove, and the cake would take far longer to cool down.

We do not have a closed system on Earth. Energy flows into our atmosphere from the sun and, to a much lesser degree, from other stars.

Materials: two bricks of identical size, heating source (such as an oven or a hot plate), container designed to retain heat (such as a pizza bag or small cooler), two surface thermometers, oven mitts

Procedure:

1. Bring the container and one brick to room temperature, if they are not already at room temperature.

2. Heat one of the two bricks in the oven at 300°F (149°C).

3. Wearing oven mitts, carefully remove the brick and check its temperature by applying a surface thermometer.

4. Insert the hot brick into the container, right next to the brick that was left at room temperature. Put a second surface thermometer on the cool brick. Seal the bag or container.

5. In one hour, check the temperatures of the two bricks. The hot one will have cooled significantly, while the cool brick will have warmed up.

6. In another hour, check again. The bricks will have the same temperatures.

Conclusion: Explain, in terms of the energy flow, what has happened to the two bricks.

As a side effect of the second law of thermodynamics, matter in a closed system is subject to **entropy,** or a tendency toward disorder. Unless energy is applied to the system, entropy always increases. An everyday example of entropy includes the mixture of gases in our atmosphere. While we can separate the gases by applying electrical energy, left to their own devices, they slip into an untidy mixture, which, at any given point on the planet, is pretty close to their overall ratio.

In the following simple experiment, you will show how disorder increases in a closed system.

Materials: tank or clear plastic box, piece of plastic long enough to reach across the width of the tank, water, blue food coloring

Procedure:

1. Fill the tank with clear water, and section the tank into two halves using the plastic barrier.

2. In one half, add blue food coloring until the water is clearly blue. Leave the other half clear.

3. Remove the barrier.

4. Observe how long it takes before the water is a uniform color. This is how long it has taken for entropy to increase.

Conclusions: Can you envision any set of circumstances in which entropy would decrease? Why or why not?

Force is a push or a pull. There are different kinds of forces; some of them are listed here:

1. the force of gravity

2. electromagnetic force

3. nuclear force

Weight is the force of gravity; it pulls down all masses. Electromagnetic force is an electrical charge that pushes or pulls and, by moving, produces magnetic fields. It is what holds atoms and molecules together. Nuclear force is millions of times stronger than gravity, but you are not aware of it because it acts only in the nucleus of an atom. It is the strongest force known.

Materials: balance, spring scale, bathroom scale

- Measure a student's weight.

- Measure the weights of several small objects around you.

- Discuss weight as the force of gravity. How does weight differ from mass?

Work is defined as a force moving through a distance. One must use energy to do work. *Energy* is the ability to do work. You can use energy without doing work. If you push a car and it does not move, you used energy but did no work. You can have motion without work. The earth rotates around the sun, yet no work is done since there is no net force pushing or pulling. The units of work have two components: the force and the distance. If you move 10 grams through a distance of 10 cm, you have work of 100 gm-cm. You can have kg-m, foot-pounds, newtons, etc. Newtons include a time component, measuring work in a unit of time. A **newton** (N) equals 1 kg moving at 1 m/sec in one second, $1 \text{ kg} \times 1 \text{ m/s/s} = 1 \text{ N}$.

Materials: spring balance, few objects, string, meterstick, 1 kg mass

- Tie one object to the spring balance with the string and pull it for a chosen distance. Multiply the force as shown on the spring balance by the distance pulled, and you have calculated the work done.

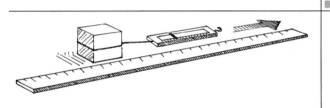

- Move a 1 kg mass over the distance of 1 m and practice until you can do it in 1 second. The force is exactly 1 newton.

Friction is a force that acts when one surface moves over another. Friction force opposes motion and changes motion (kinetic) energy into heat. Rubber heels are placed on shoes to increase friction—to prevent our slipping while walking. Friction has beneficial uses: Brakes on cars, bicycles, and other moving machines use friction to change kinetic energy into heat. On a bicycle, brake pads rub against the wheel's rim to stop the bike. You can reduce friction by rolling instead of dragging and by oiling the surfaces that are moving. Automobiles use oil to reduce the internal friction between moving parts. Bearings are used in moving machines to reduce rolling friction. Reducing friction saves work and energy. To **streamline** means to change the shape of something to offer less resistance to movement in air or water. Automobiles, planes, rockets, and ships are streamlined to offer less **drag** (friction with fluids) so that they use less fuel.

Materials: sheet of sandpaper, tabletop, small piece of carpet, mass of 500 or 1000 g, spring balance, piece of string

- Have students rub their hands together while squeezing them slightly together. The heat is kinetic energy lost to heat, somewhat welcome in cold climates.

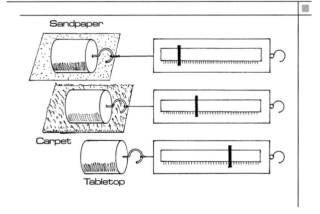

- Fasten the mass to the spring balance with the string. Drag it across the sandpaper, then the tabletop surface, and again over the carpet. Notice the values on the spring balance. The force will be greatest over the sandpaper, less over the carpet, and least over the tabletop. One surface with very little friction is ice.

The place where an object is balanced is its *center of gravity*, or *c.g.*
The object acts as if all of its matter is concentrated in this one spot.
If you stand straight up against a wall and begin to lean forward
without bending your hips, you soon will fall. An object tips over if
its center of gravity is not over its base. Objects rotate and turn most
easily about their centers of gravity.

Materials: small wooden or plastic crate, cup hook, string, rock
or several metal washers, cardboard, pin

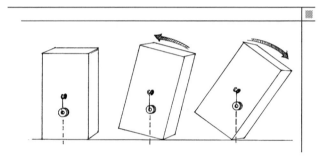

- Place the cup hook in the
geometric center of the long side
of the crate. Hang several
washers or a small rock from it.
Tilt the crate so that the washers
do not hang beyond the line of
the crate's base. The crate will
come back on its side. Tilt the
crate so that the washers hang beyond the line of its base.
The crate will fall over.

- Place the box on its broad side. Shorten the string. Show that
now it is fairly difficult to tip the crate over. The crate must be
tipped quite a lot farther than before. Measure the distances
between the c.g. and the base for both demonstrations. The
closer the c.g. is to the base, the greater the stability of the
object. Cars are hard to flip over because they have low c.g.'s
and because manufacturers widen the distance between the
tires to make their bases broader. This increases their stability.

- Cut an irregular piece of cardboard. Mark four points: *A*, *B*, *C*,
and *D*. Hang the cardboard on a wall through a pin at these
points, one at a time. While it is hanging at each point, attach
a weight to the pin with a loop of string. Trace the line of the
string on the cardboard. The four lines will converge at one
point, the center of gravity. Now hang the cardboard by its
center of gravity and spin it. The cardboard will spin freely. Try
spinning it from any of the points *A*, *B*, *C*, or *D*. The cardboard
will wobble.

Machines are devices that make work more efficient by placing forces where we want them, by moving things faster, and by making forces larger. A **lever** is a simple machine. The point where one pushes down is the **effort,** the pivot is the **fulcrum,** and the mass to be lifted is the **resistance.** The distance from the effort to the fulcrum is the *effort arm*. The distance from the resistance to the fulcrum is the *resistance arm*. Mechanical advantage is a number that tells us how many times a machine multiplies the effort used. Note that the work, done with or without a machine, is the same. (Actually, a machine absorbs some energy, so that it takes more energy to do work with a machine. However, the advantages of machines are so many that they are used extensively.) For a lever, the mechanical advantage is:

$$\text{M.A.} = \frac{\text{Length of effort arm}}{\text{Length of resistance arm}}$$

$$\text{M.A.} = \frac{200 \text{ cm}}{20 \text{ cm}} = 10$$

To find the effort, $\text{Effort} = \dfrac{\text{Resistance}}{\text{M.A.}} = \dfrac{50 \text{ kg}}{10} = 5 \text{ kg}$

Materials: stack of books, two pencils, crowbar, hammer, several long nails, piece of wood 2" × 4" × 4'

- Lift a stack of books slightly, using a pencil as a lever and another pencil at 90° to the first one as a fulcrum. With a minor effort, you will move the books. Try moving them without the pencil lever.

- Hammer a fairly large nail partway into the wood, near one end of the board. Stand on the opposite end of the wood and remove the nail—first with the hammer, then with the crowbar. The crowbar needs much less effort, for its effort arm is longer. Repeat the same activity using

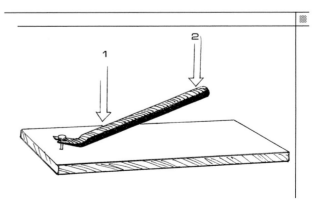

(continued)

the crowbar, first holding your hand close to the fulcrum, then as far away as possible.

- Assemble students near a door. The hinges are the fulcrum and the door-closer is the resistance. Have one or more of them try to use one finger as far away from the hinge as possible to push the door open. Next have them repeat it, moving the hand closer and closer to the hinge. As they move the one finger closer to the hinge, they have to use more and more effort to open the door, and eventually they cannot do it at all.

A **wheel and axle** make another simple machine. It is used to make work more efficient. It is a rotary lever, the lever arm being the handle of the axle. You trade turning the handle through a longer distance for the mechanical advantage. The M.A. equals the circumference of the handle's rotation divided by the circumference of the axle.

Materials: pencil sharpener mounted at the end of a table or on a board near the end of table, string, three books, a 2- to 5-kg spring balance

- Remove the cover of the sharpener. Tie the string around the books and around the center axle of the sharpener. Make sure that the string is tied tightly on the shaft of the sharpener, so that it does not slip. Start cranking up the books. The effort to raise the books will be small compared to lifting the books by hand. Use a spring balance to measure the mass of the books and then pull the handle of the sharpener with the spring balance. Compare the values.

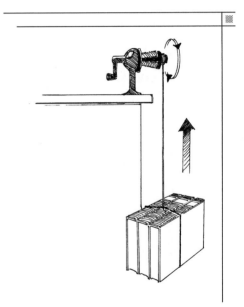

A lever is a simple machine that pivots, or completes part of a cirular motion, around a fixed point. In a first-class lever, the pivot is between the forces. Scissors, pliers, a hammer claw, a crowbar, and a teeter-totter (seesaw) are all examples of first-class levers. In teeter-totters, children adjust their balance by moving closer or farther apart from the middle (fulcrum). **Moment** or **torque** is a measure of a force's ability to rotate an object about an axis (fulcrum). The weight of children multiplied times their distance from the fulcrum on a teeter-totter is a moment. A teeter-totter works only if the moments are balanced.

Materials: two metric rulers, string, metric masses

- Place a metric ruler between two chairs or tables. Hang another metric ruler from it. Balance it by its center. With string loops, hang two different metric masses from the ruler, one on each side, so that they balance. Their moments will be the same. (The values

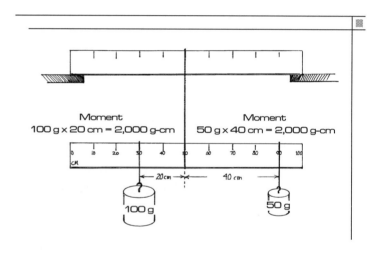

provided in the picture are merely guides.) If using paper clips or washers in place of metric masses, use large paper clips as hangers for both sides.

An inclined plane is a ramp. Steps, uphill roads, and ramps for loading and unloading barrels on trucks are other examples of ramps. The threads of a screw are also inclined planes. The mechanical advantage of a ramp is calculated with this formula:

$$M.A. = \frac{Length}{Height} = \frac{5 \text{ m}}{1 \text{ m}} = 5$$

Note that a ramp makes the rolling of a barrel more efficient, for you need only 10 kg of force. The final work is still 50 kg-m (10 kg × 5 m). The ramp helps us use less force over a greater distance.

Materials: inclined plane or board of wood, toy automobile, spring balance

- Lift the toy auto with the spring balance to measure the force needed. Place the car on an inclined plane and measure the effort needed to pull it up. The effort will be considerably less than lifting the car straight up. You are trading the lower force for a longer distance. If

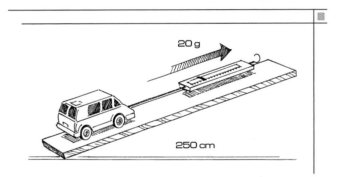

the car needs 50 g to move 100 cm, the work is 5,000 g-cm. If you wanted to pull it up a ramp using 20 g of force, you would need a ramp 250 cm long, for the work would still be 5000 g-cm.

A **pulley** is a wheel that turns on an axle. It is a simple machine that can change the direction of a force. Pulleys can be fixed or movable. A clothesline has two fixed pulleys on its ends. A block and tackle is a movable unit made with several pulleys. It is used to lift heavy objects like car engines.

Materials: two small pulleys (same size), two double pulleys (same size), string, five books of equal mass, three cup hooks, a piece of wood 2" × 4" × 3' or several lab stands

- Install the cup hooks on the board near its center and about 6 inches apart from each other. Place the board between two chairs or tables. Suspend one pulley from a hook. Pass some string through the pulley's grooves and attach a book to each end of the string. The books will not move, for they are balanced. The fulcrum is the axle and the arm is the radius of the pulley. Since both sides are equal, there is no M.A. A single pulley is used to change only the direction of force. Lift or lower a book. Both travel at the same speed.

- Attach the assembly of two pulleys. One hook will actually support two books. The top pulley is a fixed pulley; the bottom pulley is a movable pulley.

- The same activity can be extended by using four pulleys and obtaining a greater mechanical advantage. (Tackles usually have multiple pulleys side by side.) Attach the two multiple pulleys as in the illustration. Note that both upper and lower pulleys share an axle. The upper pulley system is hung by one hook, the lower one supports the books. You can also vary this activity by using spring balances to measure the force necessary to lift the books with the various pulley arrangements. This particular activity illustrates clearly the concept of mechanical advantage.

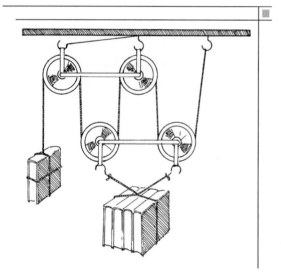

Machines are made up of many parts that move. Moving parts rub against each other, and the rubbing causes friction. Friction force not only opposes the moving force, but also changes moving energy into wasted heat. Additionally, friction causes moving parts to wear out. Since friction wastes some of the moving energy, the mechanical advantage is always less than the *ideal M.A.* The *actual M.A.* is obtained by dividing the resistance by the effort. The **efficiency** of a machine is obtained by dividing the actual M.A. by the ideal M.A.:

$$\text{Efficiency} = \frac{\text{Actual M.A.}}{\text{Ideal M.A.}}$$

The efficiency of a machine is usually expressed as a percentage. The friction in most machines is kept to the lowest possible point through oiling. In automobiles, the engines are continuously lubricated with oil. Smaller appliances with many moving parts, like sewing machines and bicycles, require constant oiling. Machines are not ideal, so they put out less work than is put into them. The difference is friction.

Materials: pulley, string, book(s), spring balance, hook, wood board

- Tie one end of the string to the hook, thread the other through the pulley and tie it to the spring balance. Using the pulley, lift up one or more books. Measure the force needed to lift the books. In the illustration, the ideal M.A. is 2. The actual M.A. $= \frac{4 \text{ kg}}{2.2 \text{ kg}} = 1.8$. The efficiency of this machine, then, is $4 \text{ kg} = \frac{1.8}{2} = 0.9 = 90\%$. In this situation, 10% of the work is lost to friction.

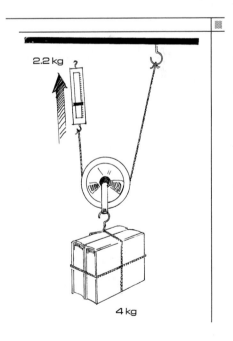

2.2 kg

4 kg

About 300 years ago, Isaac Newton discovered the laws of gravity. Gravity explains why things fall and why Earth remains in orbit around the sun, why the moon orbits Earth, and what governs the entire orderly movement of objects in the universe. Newton stated that all masses in the universe have a mutual force of attraction. This attraction increases directly with the size of their masses and decreases rapidly with the distance between them.

Newton's formula is: $F = G \times \dfrac{(M \times m)}{d^2}$

F is force on a mass.

M is one mass.

m is the other mass.

d is the distance between the masses.

G = a gravitational constant: $6.7 \times 10^{-11}\,\text{N m}^2\,\text{kg}^{-2}$.

Gravitational forces are negligible unless one of the masses is very large. Since the force depends on the inverse of the distance squared, distance is also a big factor. At 3 meters, the force is 1/9; at 5 meters it is 1/25; and at 50 meters it is 1/2,500. Since Earth has a large mass, it has a large gravitational force that attracts everything on it or near it. A space shuttle is in a free fall in space and would fall away from Earth into the depths of the universe were it not for Earth's gravitational pull, which keeps it in orbit. Astronauts adjust their orbital speed so that their momentum matches Earth's gravity. If a space shuttle were to travel faster, it would escape Earth's pull and go into deep space, as many interplanetary space probes do. If the shuttle were to slow down, it would begin to come back toward Earth.

Materials: piece of chalk, roll of masking tape, a couple of feet of string, several small objects

(continued)

- Take any small object and let it fall. Explain that all objects on or near a planet fall toward it. Throw a piece of chalk or other small object up. It will eventually fall down. As it moves up, gravity slows it down until it stops. Then it begins to free-fall.

- Take a roll of masking tape or other small object and tie it to a piece of string about a foot long. Spin it around your arm in a circular orbit. Explain to your students that this string is a visible model representing the force of gravity, where you and the string represent gravitation, and the tape represents an object in a gravitational field. If gravity ceases or the object speeds up, the object will continue falling in space.

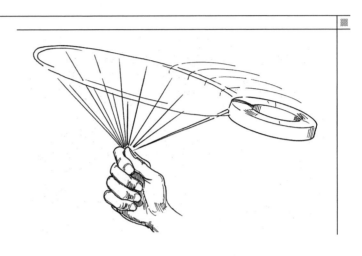

- Let go of the string and let the masking tape fly and fall somewhere where it will not harm any student.

Earth's force of gravity attracts all objects toward its theoretical center. Surveyors and contractors use plumb bobs, knowing that they will always provide a true vertical line. Air and liquids, which create friction, or drag, slow down all falling objects. Falling objects continue to accelerate (change their speed) due to the force of gravity, until their drag with the air or liquid equals their own weight. Their maximum final constant velocity is terminal velocity. In a vacuum, heavy and light objects fall at the same speed. Galileo (1564–1642), an Italian scientist, proved that all objects fall the same distance in the same time. Acceleration due to Earth's gravity is 9.8 m/s/s.

Materials: several coins of different sizes, several small objects, two sheets of paper, ruler, deck of playing cards

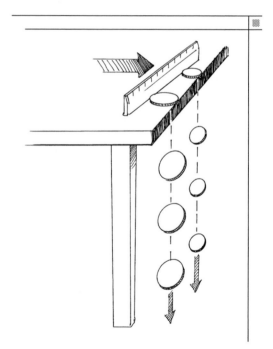

- Place two different coins near the edge of a table. Push them over the edge carefully, using the ruler, so that both start falling at the same time. Check to see if they land at the same time. They will land at the same time because the difference in their masses does not affect their acceleration.

- Crumple a piece of paper into a small, tight ball. Drop it at the same time as a flat sheet of paper. The flat piece of paper will fall more slowly, due to its greater drag with air. Modern automobiles, jets, and rockets are streamlined to reduce drag.

Mass is the amount of material in an object. The mass of an object does not change, except in one special case. Albert Einstein predicted that the mass of an object will perceptibly increase if the object travels very fast, at or near the speed of light (300,000 km/sec or 186,000 miles/sec). (If light could bend and travel around Earth at the equator, it would make 7.5 turns in 1 second.) Although the effect occurs and can be calculated at any velocity greater than rest, it begins to be truly visible when an object moves at least 80,000 km/sec. Scientists have succeeded in speeding up electrons and other small particles. They have also been able to measure the increased mass of these speeded-up particles.

Weight is the pull of gravity on an object. If an object has a mass of 1 kg, Earth's gravity pulls on it with a force of 1 kg. Different objects, even if they have the same shape and volume, can have different weights. Some materials are more dense and, therefore, heavier than others. If the force of gravity changes, so does the weight of an object. On Earth, objects appear to have less weight as they move farther from sea level. This is due to the increase in distance from Earth's center. The moon has one sixth as much gravity as Earth. An astronaut on the moon weighs only one sixth as much as on earth. In orbit, objects are weightless. This can cause serious health problems for humans traveling in space. Their muscles become flabby and they rapidly lose many vital minerals, such as calcium.

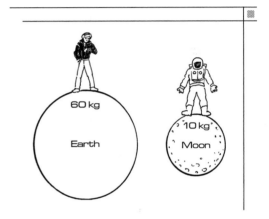

Materials: metric balance, metric mass of 500 or 1,000 g

- Place the metric mass on the balance and balance it. The balance measures weight. A 1 kg mass should have the weight of 1 kg. If it does not, calibrate the balance. (To calibrate means to adjust it so that it reads zero at the start.)

In 1905, Einstein published two papers that comprised the theory of special relativity. The papers created an almost immediate sensation. The postulates were simple, but the implications were enormous. First, Einstein did away with the idea of an absolute frame of reference—all frames of reference consider themselves "stationary," assuming they are moving at a constant velocity. Imagine playing Ping-Pong on a moving train traveling at 100 km/hr. As far as you and your partner are concerned, it's an ordinary game of Ping-Pong, just like you'd play in your basement, right? But to someone watching the game from the side of the tracks, that Ping-Pong ball is going faster than 100 km/hr. Thus, relationships between frames are relative—based on velocity differences between them. Einstein's other postulate was that the speed of light is the only thing that remains the same for all observers, wherever they may be. The speed of light is a constant.

These two ideas led Einstein to suggest that certain **relativistic effects** would be apparent to observers outside a particular frame of reference, and to suggest that nothing could exceed, or even reach, the speed of light. Observers would begin to observe relativistic effects when velocity increased well beyond the observer's rest state. Einstein used a mathematical principle to predict the amount of relativistic change an observer would notice as velocity rose, a principle called the **Lorentz transformation.**

What are the relativistic effects? Increase in mass, decrease in length, and a dilation of time. All three have been observed in laboratory settings. Objects moving at close to the speed of light gain mass, get shorter, and appear to experience time much more slowly than an observer at "rest" by a particular factor that is a function of the velocity of the object. This factor, called *gamma*, increases as velocity increases, and at close to the speed of light, it is so high that an object traveling with high velocity would appear not to be moving at all. While all objects moving faster than 0 gain gamma, the results are not easily perceptible until velocity is quite high. Today, you are going to examine the increase in mass only.

(continued)

Procedure:

While the mathematics are pretty simple, review quickly the order of operations. The order of mathematical operations can be remembered by recalling the nonsense word "PEDMAS" which stands for: Parentheses, Exponent/Radical, Division, Multiplication, Addition, and Subtraction. Division and multiplication problems are done in the order they appear in the problem, as are addition and subtraction; i.e., if you encounter a multiplication problem before a division problem, do the multiplication first.

Here is the equation for mass increase:

$$M = \frac{M_0}{\sqrt{1 - \frac{v^2}{c^2}}}$$

c is always 1, and v is velocity as a percentage of the speed of light. M_0 is whatever the rest mass of the object is.

So, if we were looking for the gamma value for an object moving at 90% of the speed of light, we would plug in .9 for velocity, and 1 for c. Let's make our lives easy and say this object has a rest mass of 100 kg. This is what the equation looks like for such an object:

$$M = \frac{100 \text{ kg}}{\sqrt{1 - \frac{9^2}{1^2}}}$$

Following the order of operations, we would first take care of the exponents: .9 squared is .81; 1 squared is 1. Then divide, and then subtract this number from 1. (1–.81 = .19):

$$M = \frac{100 \text{ kg}}{\sqrt{0.19}}$$

(continued)

We then take the square root of this number and arrive at .436:

$$M = \frac{100}{0.436}$$

Finally, divide 100 kg by .436, and discover that, at 90% of the speed of light, an object that had a rest mass of 100 kg would have a relativistic mass of 229 kg, or a gamma factor of 2.29. Quite a difference.

Conclusion: Now try a few on your own:

1. Find the relativistic mass for an object with a rest mass of 10,000 kg and a velocity of .8 of the speed of light.

2. Find the relativistic mass for an object with a rest mass of 15,000 kg and a velocity of .99 of the speed of light.

3. Find the relativistic mass for an object with a rest mass of 200,000 kg and a velocity of .75 of the speed of light.

Pressure is a force over a limited amount of surface. Some of its units are kg/cm^2, lb/in^2, or pascals, an S.I. unit. At sea level, pressure is also measured as a column of mercury 760 mm high. If you press your hand on an ice block, the block does not break. If you use an ice pick, it concentrates the force in a very tiny area, and the resulting pressure breaks the ice. If you stand on the floor, only part of your shoes is in contact with the ground. If 10 square inches of your shoes are in contact with the floor and you weigh 150 pounds, then the pressure on the floor is $15 \ lb/in^2$.

$$Pressure = \frac{Force}{Area}$$

A sharp knife has less area in contact with an object than a dull knife. Since the contact area is small, the pressing force acts as if it is increased. This is why a sharp knife cuts more easily than a dull one.

Materials: pencil, Magdeburg sphere with vacuum pump

- Have your students hold a pencil and press the tip against the palm of their other hand. They will feel a sharp point with pressure. Have them repeat this activity by turning the pencil upside down so that the eraser is against the palm. The pressure will appear to be less, for the force is concentrated over a greater area.

- Using a vacuum pump, evacuate the air from the Magdeburg sphere. Ask individual students to try to take the hemispheres apart, without opening the inlet air valve. A very large force will be needed to accomplish it. You may wish to compute the surface area of the sphere and the pressure on it.

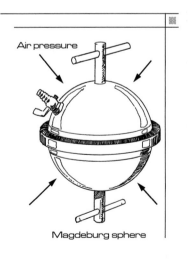

Air pressure

Magdeburg sphere

Special Safety Consideration: Should a student manage to get the sphere open, he or she will be subjected to the equal but opposite force of his exertion, which will be quite high. Make sure there are no tables, equipment, or walls in the immediate vicinity, and be sure that other students stay far away.

Pressure is measured with a **manometer.** A manometer is a U-shaped tube filled with colored water. Some manometers are filled with colored oil. Other pressure gauges are similar to aneroid barometers. Pressure gauges are used to measure tire pressure and oil pressure, among other uses in automobiles. In aviation, pressure is used to measure flight speed and for the vital fluid pressures in the aircraft and its power system. In early aviation, pressure was used to find altitudes. Pressure gauges are used on air compressors and most pressure-operated devices. Pressure gauges are also used by sea divers to measure their depths. The high pressures of oceanic depths cause nitrogen (75% of the air we breathe) to dissolve in the diver's blood system. If divers come up too rapidly, they will experience the bends, pain, and possibly death. The nitrogen will bubble in the circulatory system. To allow for the dissolved nitrogen to leave the body, divers must surface slowly. Then the nitrogen returns slowly to the lungs and is exhaled.

Materials: glass tubing about 18 inches, 3 to 4 feet of rubber hose to fit the tubing, Bunsen burner, hot-melt glue gun, glue sticks, water, food coloring, 6" × 6" board, 6" × 10" board, 1" × 1" × 12" support frame, thistle funnel or other small funnel to fit into the rubber hose, rubber sheeting, rubber band, fish tank

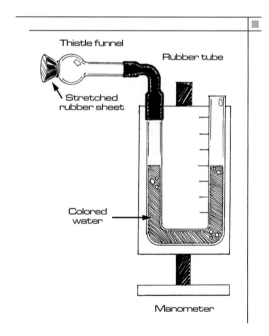

Thistle funnel

Rubber tube

Stretched rubber sheet

Colored water

Manometer

Manometer Assembly: Bend the glass tube, using a Bunsen burner. Try to obtain a shape similar to a letter U, with a 2" to 3" horizontal section. Fasten the assembly to the board with some hot-melt glue. Let the end where you will connect the rubber hose extend slightly off the board. Attach this board to the support wood with the glue and fasten the support piece to the base.

(continued)

Fasten one end of the rubber tube to the glass tubing and the other end to a thistle funnel or other small funnel. Using colored water, fill the glass tubing up to about 1 cm below the rubber tubing. Stretch the rubber membrane over the thistle funnel and fasten it with a rubber band, stretched over itself several times. Make reference marks on the opposite end of the manometer.

- Press on the rubber membrane with your thumb. You will notice a rise in pressure.

- Place the thistle funnel inside a fish tank filled with water. Show the increase of pressure with depth.

- Demonstrate that at a specific depth there is no change in pressure. Let the funnel lie on the bottom of the fish tank and turn it in several directions.

Density is a physical property of matter. It tells how much matter there is in 1 cubic centimeter. The density of water, a standard, is 1 gram per cubic centimeter. If objects of the same size have different masses, it is because they have different densities. The density of an object is found by dividing its mass by its volume. (The next activity, Archimedes' principle, addresses how to find the volume of irregular objects.)

$$\text{Density} = \frac{\text{Mass}}{\text{Volume}}$$

Following are the densities of a few familiar materials:

TABLE OF DENSITIES
(Also see Appendix.)

Material	Density g/cm^3
Gold	19.3
Mecury	13.6
Lead	11.3
Copper	8.9
Brass	8.5
Iron	7.9
Aluminum	2.7
Sugar	1.6
Corn syrup	1.1
Water	1.0
Salad oil	0.9
Wood (Elm)	0.8
Alcohol	0.8
Cork	0.2

Specific gravity is nothing more than the density of a material without its units. This is accomplished by dividing the density of

(continued)

a material by that of water. The numerical values remain the same while the units divide out.

Materials: two beakers, salad oil, white vinegar, corn syrup (dense, clear), blue and yellow food coloring, water, overhead projector

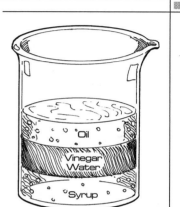

- Place a beaker on your overhead projector, for bottom illumination. Inform your students that you are about to make a salad dressing. Ask them to name the various ingredients needed. Pour a little salad oil in the beaker. Add the same amount of vinegar, to which you have added a drop of blue coloring for better visibility. Add about the same amount of corn syrup. You will obtain a beautiful separation of substances, due to their different densities. Pour the dressing back and forth a few times between the two beakers, then let it stand. The substances will separate again by their densities. Ask the class where water would go if it were added to this salad dressing. Then slowly pour in the same amount of water, after giving it a drop of yellow coloring. It will mix with the vinegar, changing the blue to green.

About 2,300 years ago, Archimedes, the Greek scientist who developed **Archimedes' principle,** lived in Syracuse, Sicily. He observed that as he floated in a bath, he appeared to be lighter, and that the level of water rose as he entered the tub. He discovered that as a body is placed in water, it displaces its own volume of water. This discovery allows us to find the volume of any irregular body by dipping it in water. Archimedes also noticed that the weight of the liquid that was displaced was the same as the apparent loss in weight of the item placed in water. This is Archimedes' principle.

Materials: overflow dish, graduated cylinder, water, small rock, spring balance, string

- Hang the rock by a string from the spring balance and measure its mass. Then submerge the rock in the water in the graduated cylinder and measure the rock's mass. Measure the volume of displaced water. Then compare the volume of displaced water to the apparent loss of mass of the rock when immersed in water. They should be nearly equal, for 1 cm^3 of water has a mass of 1 g.

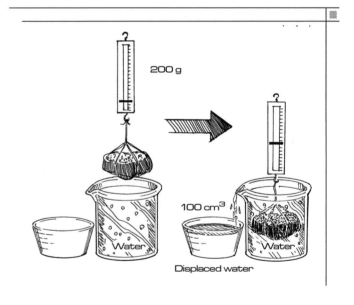

200 g

100 cm^3

Displaced water

Applied Conclusion: When you put an object in water and it appears to become lighter, the difference between its mass in air and in water represents its volume in cm^3. No overflow dish is really needed, except for the original proof.

Sound is caused by objects vibrating back and forth. With a tuning fork, you can make a sound of only one pitch (frequency).

Materials: tuning fork, glass nearly full of water, rubber mallet if available, record player, old record, index card, sheet of paper, two metal pins

- Hit the tuning fork with the rubber mallet. If a mallet is not available, hit the tuning fork on the rubber heel of your shoe. Hitting the fork on a hard material causes the fork to lose its calibrated pitch (frequency). Point the fork toward the class and repeat several times.

- Touch the handle of the vibrating fork to a student desk and have the students listen by placing their ears on the surface. Repeat for the entire class.

- Since it may be difficult to see that the tuning fork is really vibrating, place the vibrating fork into the glass of water. It will create a gentle spray. The water acts as a dampener (reducer) of energy. The tuning fork will quickly cease to vibrate. The water behaves like a car's shock absorbers or a piano's dampener. In each case, kinetic energy is transferred into another material by doing work (pushing water or oil or piano strings).

- Place the record on the record player. Make a player needle by pushing one pin through an index card. Holding the card gently, place the needle in the groove of the record. You will be able to hear a faint sound from the record as the card vibrates in tune with the needle. Thomas Alva Edison invented the phonograph, the "talk-ing machine" of his day. If you need more sound, make a paper cone, fold over the closed end, and place a pin through it. Again, place the needle in the record groove. Louder sound will be heard. The paper acts as a mechanical amplifier.

Decibels are units of measurement for the loudness of sounds. They are logarithmic; this means that they are not linear. Any change of ±3 dB means a doubling or halving of the loudness of sound. Since sound is transmitted through material objects as well as air, decibels measure the pressure of the sound waves. Since all sounds are made by vibrating objects, those that have more energy appear to be louder. When a person is exposed to loud sounds, the eardrum protects itself by growing thicker, an irreversible process known as deafness. People exposed to loud sounds must wear protective ear devices. Exposure to 90 dB can cause temporary deafness. Electronic supply stores carry inexpensive yet accurate decibel meters, which you can use to check the level of sounds to which you are exposed. Use one to check sound around you and during school events like dances.

TABLE OF SOUND IN DECIBELS

Sound Pressure Decibels	Sound Source
0	No sound
15	Rustling of leaves
40	Whisper
45	Normal household
65	Normal talk
75	Busy street traffic
85	Vacuum cleaner
100	Subway or riveting
110	Thunder
115	Rock band near
120	Painful to ears
160	Jet engine
200	Rocket takeoff

(continued)

Materials: toy xylophone, whistle, tuning fork, radio, other sound makers

- Play the various sound-making devices. Finally, play anything on the radio and adjust the volume from very low to fairly loud and back down again.

Materials: tuning fork, **cathode ray oscilloscope (CRO),** microphone, radio

- Hook up the microphone to the CRO. Show the waves of sound caused by the tuning fork. Show the waves of a faint music program. Repeat with a greater volume. The waves will appear to be higher. The height of a wave is its amplitude, and it is the measure of a wave's energy—in the case of sound, its loudness.

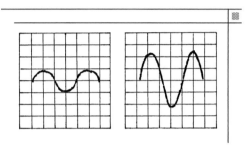

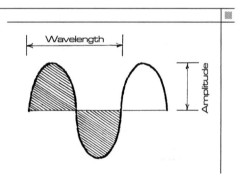

Vibrations that cause sound are between 20 and 20,000 vibrations per second. This range is called the *audio spectrum*. The number of vibrations per second is **frequency.** The unit of frequency is the **hertz (Hz),** after Heinrich Hertz, who did much pioneering work that led to the invention of radio. Sounds above the audio spectrum, too high to be heard by humans, can be heard by dogs and many other animals. This explains why certain dog whistles are soundless to humans.

Materials: three tuning forks of different frequencies, small radio with bass and treble adjustments, sine-square wave generator (optional), speaker, cathode ray oscilloscope (CRO)

- Play each tuning fork separately, starting from the lowest and going to the highest. Point out that as you increase the pitch (frequency), the sound gets higher in frequency. Each fork will have stamped on its body its vibrating frequency.

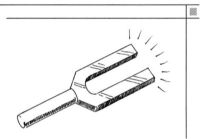

- Play some music on the radio. Turn the bass control up and down, then repeat with the treble control. Students will immediately comprehend higher and lower pitch.

- Connect the sound generator to the speaker and to the CRO. Go from lower- to higher-pitched sounds. Students will not only hear sound differences clearly, but will also be able to observe how the number of waves increases as you increase frequency.

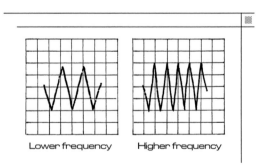

Lower frequency   Higher frequency

Easy Science Demos & Labs:
Physics

Sound is transmitted through material objects (solids, liquids, and gases) because sound waves cause compression and rarefaction (expansion) in the distance between the molecules of matter. Where there is no matter, as in a vacuum, sound is not transmitted. Sound eventually stops, because its energy is used up in moving the molecules of matter. Sound is transmitted by either **compression** or **transverse waves.** The speed of sound is about 335 meters per second in air, 1,500 meters per second in water, and about 5,000 meters per second in metals. Since sound travels at a fraction of the speed of light, if you count the time between seeing a bolt of lightning and hearing the thunder, you will know how far away the lightning has struck. Sound travels 1 kilometer in 3 seconds.

Materials: Slinky, tuning fork, glass of water

- Have one student hold one end of the Slinky about 20 feet away from you. Shake it up and down and show the travel of a **transverse wave.** You may wish to place it on a flat table and shake one end. If you shake while a wave is returning, the two waves will collide. Waves can aid (help each other) or buck (cancel their energies).

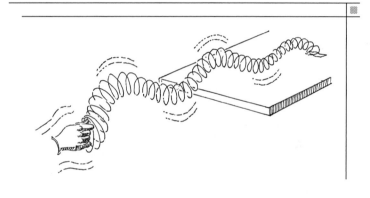

- While holding the Slinky stretched, push rapidly along the length of the spring, not laterally or up and down. You will notice that there is a small region of compression of springs that travels forth and back. That is a **compression** wave. A small amount of transverse wave is unavoidable. These two activities will work equally well on Earth and in space.

- Generate a transverse wave and hold one end of the Slinky to a student's ear. Pluck the Slinky. It will make many interesting sounds.

Most sounds are made up of a combination of several frequencies. A tuning fork makes a pure sound, for it produces only one frequency. The particular frequency at which an object vibrates is its **natural frequency.** When you have two objects with the same natural frequency, one will vibrate when the other one does. This phenomenon is **resonance.** Resonance explains all radio transmissions. Your radio has an electronic tuning fork that adjusts to the frequencies of the transmitting stations. As you tune in a particular frequency, the radio station signal is heard.

Materials: two boxes (open on one end) with identical tuning forks mounted on top, rubber mallet (resonance apparatus)

- Place the openings of the boxes so that they face each other a few inches apart. Hit one tuning fork. Observe how the other one responds by vibrating. Since a small amount of energy will be transmitted, 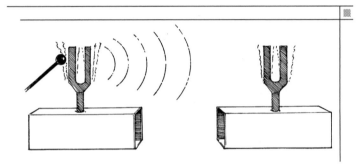 have students listen by placing the ear on the edge of the box. Change the frequency of the fork by sliding the adjustable sleeve up or down. Hit one fork and adjust the other fork until it is tuned to the first one.

Appendix

1. Assessing Laboratory Reports

This book contains 10 student laboratory assignments, for which students should be expected to produce written reports. Go over what you want to see in a lab report with your students before they start. Information should include:

- **Purpose**: Why is this lab being performed? What is the objective of the lab?

- **Hypothesis:** Given the initial level of knowledge, what do students expect for an outcome and why?

- **Materials list:** Students should be told that one of the main reasons for writing lab reports is so that others can replicate the lab. A well-organized materials list makes it easier for a reader to understand the lab, and makes redoing the experiment much easier as well.

- **Procedure:** Likewise, a student should include each step of the procedure that the lab partners or group actually followed.

- **Data:** What events or measurements were observed in the lab? In physics, observations of changes in states of matter, as well as quantifications of findings, are all very important.

- **Conclusion:** What were the results? What were the limitations? Did a student's hypothesis match the data? If something went wrong, what does the student think happened?

(continued)

1. Assessing Laboratory Reports (continued)

In order to give you a quick guide to assessing lab reports, we have constructed the following rubric:

	1	2	3	4
Understanding of Concept	Poor	Adequate	Good	Outstanding
Methodology	Poor	Adequate	Good	Outstanding
Organization of Experiment	Poor	Adequate	Good	Outstanding
Organization of Report	Poor	Adequate	Good	Outstanding

Laboratory reports are an important stepping stone for young scientists, but can become burdensome to correct. We hope this rubric assists the typical busy teacher in providing a quality lab experience without sacrificing the deeper knowledge of the scientific method that writing lab reports reinforces among young scientists.

2. Temperature Conversion (Celsius to Fahrenheit)

C°	F°	C°	F°	C°	F°	C°	F°	C°	F°	C°	F°
250	482.00	200	392.00	150	302.00	100	212.00	50	122.00	0	32.00
249	480.20	199	390.20	149	300.20	99	210.20	49	120.20	−1	30.20
248	478.40	198	388.40	148	298.40	98	208.40	48	118.40	−2	28.40
247	476.60	197	386.60	147	296.60	97	206.60	47	116.60	−3	26.60
246	474.80	196	384.80	146	294.80	96	204.80	46	114.80	−4	24.80
245	473.00	195	383.00	145	293.00	95	203.00	45	113.00	−5	23.00
244	471.20	194	381.20	144	291.20	94	201.20	44	111.20	−6	21.20
243	469.40	193	379.40	143	289.40	93	199.40	43	109.40	−7	19.40
242	467.60	192	377.60	142	287.60	92	197.60	42	107.60	−8	17.60
241	465.80	191	375.80	141	285.80	91	195.80	41	105.80	−9	15.80
240	464.00	190	374.00	140	284.00	90	194.00	40	104.00	−10	14.00
239	462.20	189	372.20	139	282.20	89	192.20	39	102.20	−11	12.20
238	460.40	188	370.40	138	280.40	88	190.40	38	100.40	−12	10.40
237	458.60	187	368.60	137	278.60	87	188.60	37	98.60	−13	8.60
236	456.80	186	366.80	136	276.80	86	186.80	36	96.80	−14	6.80
235	455.00	185	365.00	135	275.00	85	185.00	35	95.00	−15	5.00
234	453.20	184	363.20	134	273.20	84	183.20	34	93.20	−16	3.20
233	451.40	183	361.40	133	271.40	83	181.40	33	91.40	−17	1.40
232	449.60	182	359.60	132	269.60	82	179.60	32	89.60	−18	−0.40
231	447.80	181	357.80	131	267.80	81	177.80	31	87.80	−19	−2.20
230	446.00	180	356.00	130	266.00	80	176.00	30	86.00	−20	−4.00
229	444.20	179	354.20	129	264.20	79	174.20	29	84.20	−21	−5.80
228	442.40	178	352.40	128	262.40	78	172.40	28	82.40	−22	−7.60
227	440.60	177	350.60	127	260.60	77	170.60	27	80.60	−23	−9.40
226	438.80	176	348.80	126	258.80	76	168.80	26	78.80	−24	−11.20
225	437.00	175	347.00	125	257.00	75	167.00	25	77.00	−25	−13.00
224	435.20	174	345.20	124	255.20	74	165.20	24	75.20	−26	−14.80
223	433.40	173	343.40	123	253.40	73	163.40	23	73.40	−27	−16.60
222	431.60	172	341.60	122	251.60	72	161.60	22	71.60	−28	−18.40
221	429.80	171	339.80	121	249.80	71	159.80	21	69.80	−29	−20.20
220	428.00	170	338.00	120	248.00	70	158.00	20	68.00	−30	−22.00
219	426.20	169	336.20	119	246.20	69	156.20	19	66.20	−31	−23.80
218	424.40	168	334.40	118	244.40	68	154.40	18	64.40	−32	−25.60
217	422.60	167	332.60	117	242.60	67	152.60	17	62.60	−33	−27.40
216	420.80	166	330.80	116	240.80	66	150.80	16	60.80	−34	−29.20
215	419.00	165	329.00	115	239.00	65	149.00	15	59.00	−35	−31.00
214	417.20	164	327.20	114	237.20	64	147.20	14	57.20	−36	−32.80
213	415.40	163	325.40	113	235.40	63	145.40	13	55.40	−37	−34.60
212	413.60	162	323.60	112	233.60	62	143.60	12	53.60	−38	−36.40
211	411.80	161	321.80	111	231.80	61	141.80	11	51.80	−39	−38.20
210	410.00	160	320.00	110	230.00	60	140.00	10	50.00	−40	−40.00
209	408.20	159	318.20	109	228.20	59	138.20	9	48.20	−41	−41.80
208	406.40	158	316.40	108	226.40	58	136.40	8	46.40	−42	−43.60
207	404.60	157	314.60	107	224.60	57	134.60	7	44.60	−43	−45.40
206	402.80	156	312.80	106	222.80	56	132.80	6	42.80	−44	−47.20
205	401.00	155	311.00	105	221.00	55	131.00	5	41.00	−45	−49.00
204	399.20	154	309.20	104	219.20	54	129.20	4	39.20	−46	−50.80
203	397.40	153	307.40	103	217.40	53	127.40	3	37.40	−47	−52.60
202	395.60	152	305.60	102	215.60	52	125.60	2	35.60	−48	−54.40
201	393.80	151	303.80	101	213.80	51	123.80	1	33.80	−49	−56.20

3. Temperature Conversion (Fahrenheit to Celsius)

F°	C°	F°	C°	F°	C°	F°	C°	F°	C°	F°	C°
250	121.11	200	93.33	150	65.56	100	37.78	50	10.00	0	−17.78
249	120.56	199	92.78	149	65.00	99	37.22	49	9.44	−1	−18.33
248	120.00	198	92.22	148	64.44	98	36.67	48	8.89	−2	−18.89
247	119.44	197	91.67	147	63.89	97	36.11	47	8.33	−3	−19.44
246	118.89	196	91.11	146	63.33	96	35.56	46	7.78	−4	−20.00
245	118.33	195	90.56	145	62.78	95	35.00	45	7.22	−5	−20.55
244	117.78	194	90.00	144	62.22	94	34.44	44	6.67	−6	−21.11
243	117.22	193	89.44	143	61.67	93	33.89	43	6.11	−7	−21.67
242	116.67	192	88.89	142	61.11	92	33.33	42	5.56	−8	−22.22
241	116.11	191	88.33	141	60.56	91	32.78	41	5.00	−9	−22.78
240	115.56	190	87.78	140	60.00	90	32.22	40	4.44	−10	−23.33
239	115.00	189	87.22	139	59.44	89	31.67	39	3.89	−11	−23.89
238	114.44	188	86.67	138	58.89	88	31.11	38	3.33	−12	−24.44
237	113.89	187	86.11	137	58.33	87	30.56	37	2.78	−13	−25.00
236	113.33	186	85.56	136	57.78	86	30.00	36	2.22	−14	−25.56
235	112.78	185	85.00	135	57.22	85	29.44	35	1.67	−15	−26.11
234	112.22	184	84.44	134	56.67	84	28.89	34	1.11	−16	−26.67
233	111.67	183	83.89	133	56.11	83	28.33	33	0.56	−17	−27.22
232	111.11	182	83.33	132	55.56	82	27.78	32	0.00	−18	−27.78
231	110.56	181	82.78	131	55.00	81	27.22	31	−0.56	−19	−28.33
230	100.00	180	82.22	130	54.44	80	26.67	30	−1.11	−20	−28.89
229	109.44	179	81.67	129	53.89	79	26.11	29	−1.67	−21	−29.44
228	108.89	178	81.11	128	53.33	78	25.56	28	−2.22	−22	−30.00
227	108.33	177	80.56	127	52.78	77	25.00	27	−2.78	−23	−30.56
226	107.78	176	80.00	126	52.22	76	24.44	26	−3.33	−24	−31.11
225	107.22	175	79.44	125	51.67	75	23.89	25	−3.89	−25	−31.67
224	106.67	174	78.89	124	51.11	74	23.33	24	−4.44	−26	−32.22
223	106.11	173	78.33	123	50.56	73	22.78	23	−5.00	−27	−32.78
222	105.56	172	77.78	122	50.00	72	22.22	22	−5.56	−28	−33.33
221	105.00	171	77.22	121	49.44	71	21.67	21	−6.11	−29	−33.89
220	104.44	170	76.67	120	48.89	70	21.11	20	−6.67	−30	−34.44
219	103.89	169	76.11	119	48.33	69	20.56	19	−7.22	−31	−35.00
218	103.33	168	75.56	118	47.78	68	20.00	18	−7.78	−32	−35.56
217	102.78	167	75.00	117	47.22	67	19.44	17	−8.33	−33	−36.11
216	102.22	166	74.44	116	46.67	66	18.89	16	−8.89	−34	−36.67
215	101.67	165	73.89	115	46.11	65	18.33	15	−9.44	−35	−37.22
214	101.11	164	73.33	114	45.56	64	17.78	14	−10.00	−36	−37.78
213	100.56	163	72.78	113	45.00	63	17.22	13	−10.56	−37	−38.33
212	100.00	162	72.22	112	44.44	62	16.67	12	−11.11	−38	−38.89
211	99.44	161	71.67	111	43.89	61	16.11	11	−11.67	−39	−39.44
210	98.89	160	71.11	110	43.33	60	15.56	10	−12.22	−40	−40.00
209	98.33	159	70.56	109	42.78	59	15.00	9	−12.78	−41	−40.56
208	97.78	158	70.00	108	42.22	58	14.44	8	−13.33	−42	−41.11
207	97.22	157	69.44	107	41.67	57	13.89	7	−13.89	−43	−41.67
206	96.67	156	68.89	106	41.11	56	13.33	6	−14.44	−44	−42.22
205	96.11	155	68.33	105	40.56	55	12.78	5	−15.00	−45	−42.78
204	95.56	154	67.78	104	40.00	54	12.22	4	−15.56	−46	−43.33
203	95.00	153	67.22	103	39.44	53	11.67	3	−16.11	−47	−43.89
202	94.44	152	66.67	102	38.89	52	11.11	2	−16.67	−48	−44.44
201	93.89	151	66.11	101	38.33	51	10.56	1	−17.22	−49	−45.00

4. Melting and Boiling Points of Elements

ELEMENT	Sym	Melting Point °C	Boiling Point °C	ELEMENT	Sym	Melting Point °C	Boiling Point °C
Actinium	Ac	1050	3200 ± 300	Mercury	Hg	−38.87	356.58
Aluminum	Al	0660.37	2467	Molybdenum	Mo	2617	4612
Americium	Am	994 ± 4	2607	Neodymium	Nd	1021	3068
Antimony	Sb	630.74	1750	Neon	Ne	−248.67	−246.048
Argon	Ar	−189.2	−185.7	Neptunium	Np	640 ± 1	3902
Arsenic (gray)	As	817 (28 atm)	613 (sub)	Nickel	Ni	1453	2732
Astatine	At	302	337	Niobium-Columbium	Nb	2468 ± 10	4742
Barium	Ba	725	1640	Nitrogen	N	−209.86	−195.8
Berkelium	Bk	–	–	Nobelium	No		
Beryllium	Be	1278 ± 5	2970 (5 mm)	Osmium	Os	3045 ± 30	5027 ± 100
Bismuth	Bi	271.3	1560 ± 5	Oxygen	O	−218.4	−182.962
Boron	B	2300	2550 (sub)	Palladium	Pd	1552	3140
Bromine	Br	−7.2	58.78	Phosphorus	P	44.1 (white)	280 (white)
Cadmium	Cd	320.9	765	Platinum	Pt	1772	3827 ± 100
Calcium	Ca	839 ± 2	1484	Plutonium	Pu	641	3232
Californium	Cf	–	–	Polonium	Po	254	962
Carbon	C	~ 3550	4827	Potassium	K	63.65	774
Cerium	Ce	799 ± 3	3426	Praseodymium	Pr	931	3512
Cesium	Cs	28.40 ± 0.01	678.4	Promethium	Pm	~ 1080	2460(?)
Chlorine	Cl	−100.98	−34.6	Protactinium	Pa	< 1600	–
Chromium	Cr	1857 ± 20	2672	Radium	Ra	700	1140
Cobalt	Co	1495	2870	Radon	Rn	−71	−61.8
Copper	Cu	1083.4 ± 0.2	2567	Rhenium	Re	3180	5627 (est)
Curium	Cm	1340 ± 40	–	Rhodium	Rh	1966 ± 3	3727 ± 100
Dysprosium	Dy	1412	2562	Rubidium	Rb	38.89	688
Einsteinium	Es	–	–	Ruthenium	Ru	2310	3900
Erbium	Er	1529	2863	Samarium	Sm	1077	1791
Europium	Eu	822	1597	Scandium	Sc	1541	2831
Fermium	Fm	–	–	Selenium	Se	217	684.9 ± 1.0
Fluorine	F	−219.62	−188.14	Silicon	Si	1410	2355
Francium	Fr	(27)	(677)	Silver	Ag	961.93	2212
Gadolinium	Gd	1313 ± 1	3266	Sodium	Na	97.81 ± 0.03	882.9
Gallium	Ga	29.78	2403	Strontium	Sr	769	1384
Germanium	Ge	937.4	2830	Sulfur	S	112.8	444.674
Gold	Au	1064.43	2807	Tantalum	Ta	2996	5425 ± 100
Hafnium	Hf	2227 ± 20	4602	Technetium	Tc	2172	4877
Helium	He	−272.2	−268.934	Tellurium	Te	449.5 ± 0.3	989.8 ± 3.8
Holmium	Ho	1474	2695	Terbium	Tb	1356	3123
Hydrogen	H	−259.14	−252.87	Thallium	Tl	303.5	1457 ± 10
Indium	In	156.61	2080	Thorium	Th	1750	~4790
Iodine	I	113.5	184.35	Thulium	Tm	1545 ± 15	1947
Iridium	Ir	2410	4130	Tin	Sn	231.9681	2270
Iron	Fe	1535	2750	Titanium	Ti	1660 ± 10	3287
Krypton	Kr	−156.6	−152.30 ± 0.10	Tungsten	W	3410 ± 20	5660
Lanthanum	La	921 ± 5	3457	Uranium	U	1132.3 ± 0.8	3818
Lawrencium	Lr	–	–	Vanadium	V	1890 ± 10	3380
Lead	Pb	327.502	1740	Wolfram	(see	Tungsten)	
Lithium	Li	180.54	1347	Xenon	Xe	−111.9	−107.1 ± 3
Lutetium	Lu	1663 ± 5	3395	Ytterbium	Yb	819	1194
Magnesium	Mg	648.8 ± 0.5	1090	Yttrium	Y	522	3338
Manganese	Mn	1244 ± 3	1962	Zinc	Zn	19.58	907
Mendelevium	Md	–	–	Zirconium	Zr	1852 ± 2	4377

sub = sublimation

5. Range of Resistances

	ohm-meters at 0°C
Aluminum	2.63×10^{-8}
Amber	5×10^{14}
Boron	1×10^{6}
Constantin	49×10^{-8}
Copper	1.54×10^{-8}
Germanium	2×10
Glass	$10^{11} - 10^{13}$
Gold	2.27×10^{-8}
Iron	11.0×10^{-8}
Mercury	94×10^{-8}
Mica (colorless)	2×10^{15}
Nichrome	3.5×10^{-5}
Platinum	11.0×10^{-8}
Quartz (fused)	5×10^{17}
Silicon	3×10^{4}
Silver	1.52×10^{-8}
Sulfur	1×10^{15}
Tungsten	5.0×10^{-8}
Wood (maple)	3×10^{8}

These are only approximations since exact values vary.

6. Coefficients of Volume Expansion

(Increase per unit volume per $°C \times 10^{-3}$)

Alcohol (ethyl)	1.1
Benzene	1.2
Carbon tetrachloride	1.2
Ether	1.7
Gases (at 0°C and constant pressure)	3.7
Glass	0.025
Glycerin	0.5
Mercury	0.18
Petroleum	0.96
Turpentine	0.97
Water (at 20°C)	0.21

7. Electrochemical Equivalents

Element and valence	Symbol and charge on ion (valence)	Grams per Coulomb or ampere-second
Aluminum	Al+++	9.32×10^{-5}
Copper (2)	Cu++	32.9×10^{-5}
Copper (1)	Cu+	65.9×10^{-5}
Gold (3)	Au+++	68.1×10^{-5}
Gold (1)	Au+	204.4×10^{-5}
Hydrogen	H+	1.04×10^{-5}
Iron (3)	Fe+++	19.3×10^{-5}
Iron (2)	Fe++	28.9×10^{-5}
Lead (4)	Pb++++	53.7×10^{-5}
Lead (2)	Pb++	107.4×10^{-5}
Nickel	NI++	30.4×10^{-5}
Oxygen	O--	8.29×10^{-5}
Silver	Ag+	111.8×10^{-5}
Zinc	Zn++	33.9×10^{-5}

8. Wavelengths of Various Radiations

Radiation	Ångström units
Cosmic rays	0.005
Gamma rays	0.005–1.40
X-rays	0.1–100
Ultraviolet	2920–4000
Visible spectrum	4000–700
Violet	4000–4240
Blue	4240–4912
Green	4912–5750
Yellow	5750–5850
Orange	5850–6470
Red	6470–7000
Maximum visibility	5560
Infrared	over 7000
Ång. to inch multiply by	3.937×10^{-9}
Ång. to cm. multiply by	1×10^{-8}

Easy Science Demos & Labs:
Physics

9. Specific Heat of Materials

(calories/gram/°C from 20–100°C)

Air (constant pressure)	0.24
Alcohol	0.66
Aluminum	0.22
Cellulose (dry)	0.37
Charcoal (at 10°C)	0.16
Clay (dry)	0.22
Copper	0.09
Glycerin	0.57
Granite	0.19
Gold	0.03
Glass (normal thermometer)	0.20
Ice (at –10°C)	0.53
Iron (cast)	0.119
Iron (wrought)	0.115
Lead	0.031
Marble	0.21
Mercury	0.033
Paraffin	0.7
Platinum	0.03
Petroleum	0.51
Silver	0.06
Water	1.00
Water vapor (constant pressure)	0.46
Water vapor (constant volume)	0.36

10. Coefficient of Linear Expansion

(approximate expansion per unit length per 0°C $\times 10^{-6}$)

Aluminum	24
Brass	19
Bronze	27
Cement	10–14
Copper	1.7
German silver	1.8
Glass (plate)	0.9
Gold	13
Granite	8
Ice	51
Iron (wrought)	11
Lead	29
Magnesium	27
Platinum	0.9
Pyrex	3.6
Quartz (fused)	0.4
Rubber (hard)	80
Silver	1.7
Steel	13
Steel (stainless)	18
Tin	27
Wood (parallel to grain) beech	0.2
elm	0.6
oak	0.5
pine	0.5
walnut	0.7
Wood (across grain) beech	61
elm	44
pine	34
walnut	48

11. Density of Liquids

approx. gm/cm^3 at 20°C

Acetone	0.79
Alcohol (ethyl)	0.79
Alcohol (methyl)	0.81
Benzene	0.90
Carbon disulfide	1.29
Carbon tetrachloride	1.56
Chloroform	1.50
Ether	0.74
Gasoline	0.68
Glycerin	1.26
Kerosene	0.82
Linseed oil (boiled)	0.94
Mercury	13.6
Milk	1.03
Naphtha (petroleum)	0.67
Olive oil	0.92
Sulfuric acid	1.82
Turpentine	0.87
Water 0° C	0.99
Water 4° C	1.00
Water–sea	1.03

12. Altitude, Barometer, and Boiling Point

altitude (approx. ft)	barometer reading (cm of mercury)	boiling point (°C)
15,430	43.1	84.9
10,320	52.0	89.8
6,190	60.5	93.8
5,510	62.0	94.4
5,060	63.1	94.9
4,500	64.4	95.4
3,950	65.7	96.0
3,500	66.8	96.4
3,060	67.9	96.9
2,400	69.6	97.6
2,060	70.4	97.9
1,520	71.8	98.5
970	73.3	99.0
530	74.5	99.5
0	76.0	100.0
−550	77.5	100.5

13. Specific Gravity

gram/cm^3 at 20°C

Agate	2.5–2.6	Granite*	2.7	Polystyrene	1.06
Aluminum	2.7	Graphite	2.2	Quartz	2.6
Brass*	8.5	Human body–normal	1.07	Rock salt	2.1–2.2
Butter	0.86	Human body–lungs full	1.00	Rubber (gum)	0.92
Cellural cellulose acetate	0.75	Ice	0.92	Silver	10.5
Celluloid	1.4	Iron (cast)*	7.9	Steel	7.8
Cement*	2.8	Lead	11.3	Sulfur (roll)	2.0
Coal (anthracite)*	1.5	Limestone	2.7	Tin	7.3
Coal (bituminous)*	1.3	Magnesium	1.74	Tungsten	18.8
Copper	8.9	Marble*	2.7	Wood: Rock Elm	0.76
Cork	0.22–0.26	Nickel	8.8	Balsa	0.16
Diamond	3.1–3.5	Opal	2.1–2.3	Red Oak	0.67
German Silver	8.4	Osmium	22.5	Southern Pine	0.56
Glass (common)	2.5	Paraffin	0.9	White Pine	0.4
Gold	19.3	Platinum	21.4	Zinc	7.1

*Non-homogeneous material. Specific gravity may vary. Table gives average value.

14. Units: Conversions and Constants

From	To	× By
Acres	Square feet	43,560
Acres	Square meters	4,046.8564
Acre-feet	Cubic feet	43,560
Avogadro's number	6.02252×10^{23}	
Barrel (US dry)	Barrel (US liquid)	0.96969
Barrel (US liquid)	Barrel (US dry)	1.03125
Bars	Atmospheres	0.98692
Bars	Grams/square centimeter	1,019.716
Cubic feet	Acre-feet	2.2956841×10^{-5}
Cubic feet	Cubic centimeters	28,316.847
Cubic feet	Cubic meters	0.028316984
Cubic feet	Gallons (US liquid)	7.4805195
Cubic feet	Quarts (US liquid)	29.922078
Cubic inches	Cubic centimeters	16.38706
Cubic inches	Cubic feet	0.0005787037
Cubic inches	Gallons (US liquid)	0.004329004
Cubic inches	Liters	0.016387064
Cubic inches	Ounces (US liquid)	0.5541125
Cubic inches	Quarts (US liquid)	0.03463203
Cubic meters	Acre-feet	0.0008107131
Cubic meters	Barrels (US liquid)	8.386414
Cubic meters	Cubic feet	35.314667
Cubic meters	Gallons (US liquid)	264.17205
Cubic meters	Quarts (US liquid)	1,056.6882
Cubic yards	Cubic centimeters	764,554.86
Cubic yards	Cubic feet	27
Cubic yards	Cubic inches	46,656
Cubic yards	Liters	764.55486
Cubic yards	Quarts (US liquid)	807.89610
Days (mean solar)	Days (sidereal)	1.0027379
Days (mean solar)	Hours (mean solar)	24
Days (mean solar)	Hours (sidereal)	24.065710
Days (mean solar)	Years (calendar)	0.002739726
Days (mean solar)	Years (sidereal)	0.0027378031
Days (mean solar)	Years (tropical)	0.0027379093
Days (sidereal)	Days (mean solar)	0.99726957
Days (sidereal)	Hours (mean solar)	23.93447
Days (sidereal)	Hours (sidereal)	24
Days (sidereal)	Minutes (mean solar)	1,436.0682

(continued)

From	To	× By
Days (sidereal)	Minutes (sidereal)	1,440
Days (sidereal)	Seconds (sidereal)	86,400
Days (sidereal)	Years (calendar)	0.0027322454
Days (sidereal)	Years (sidereal)	0.0027303277
Days (sidereal)	Years (tropical)	0.0027304336
Decibels	Bels	0.1
Decimeters	Feet	0.32808399
Decimeters	Inches	3.9370079
Decimeters	Meters	0.1
Degrees	Minutes	60
Degrees	Radians	0.017453293
Degrees	Seconds	3,600
Degrees	Circles	0.0027777
Degrees	Quadrants	0.0111111
Dekaliters	Pecks (US)	1.135136
Dekaliters	Pints (US dry)	19.16217
Dekameters	Feet	32.808399
Dekameters	Inches	393.70079
Dekameters	Yards	10.93613
Dekameters	Centimeters	1,000
Fathoms	Centimeters	182.88
Fathoms	Feet	6
Fathoms	Inches	72
Fathoms	Meters	1.8288
Fathoms	Miles (nautical International)	0.00098747300
Fathoms	Miles (statute)	0.001136363
Fathoms	Yards	2
Feet	Centimeters	30.48
Feet	Fathoms	0.166666
Feet	Furlongs	0.00151515
Feet	Inches	12
Feet	Meters	0.3048
Feet	Microns	304800
Feet	Miles (nautical International)	0.00016457883
Feet	Miles (statute)	0.000189393
Feet	Rods	0.060606
Feet	Yards	0.333333
Gallons (US liquid)	Acre-feet	3.0688833×10^{-6}
Gallons (US liquid)	Barrels (US liquid)	0.031746032

(continued)

14. Units: Conversions and Constants (continued)

From	To	× By
Gallons (US liquid)	Bushels (US)	0.10742088
Gallons (US liquid)	Cubic centimeters	3,785.4118
Gallons (US liquid)	Cubic feet	0.133680555
Gallons (US liquid)	Cubic inches	231
Gallons (US liquid)	Cubic meters	0.0037854118
Gallons (US liquid)	Cubic yards	0.0049511317
Gallons (US liquid)	Gallons (US dry)	0.85936701
Gallons (US liquid)	Gallons (wine)	1
Gallons (US liquid)	Gills (US)	32
Gallons (US liquid)	Liters	3.7854118
Gallons (US liquid)	Ounces (US fluid)	128
Gallons (US liquid)	Pints (US liquid)	8
Gallons (US liquid)	Quarts (US liquid)	4
Grains	Carats (metric)	0.32399455
Grains	Drams (apoth. or troy)	0.016666
Grains	Drams (avdp.)	0.036671429
Grains	Grams	0.06479891
Grains	Milligrams	64.79891
Grains	Ounces (apoth. or troy)	0.0020833
Grains	Ounces (avdp.)	0.0022857143
Grams	Carats (metric)	5
Grams	Drams (apoth. or troy)	0.25720597
Grams	Drams (avdp.)	0.56438339
Grams	Dynes	980.665
Grams	Grains	15.432358
Grams	Ounces (apoth. or troy)	0.032150737
Grams	Ounces (avdp.)	0.035273962
Gravitational constant	Centimeters/(second × second)	980.621
Gravitational constant = G	Dyne cm^2 g^{-2}	6.6732 (31) × 10^{-8}
Gravitational constant	Feet/(second × second)	32.1725
Gravitational constant = G	N m^2 kg^{-2}	6.6732 (31) × 10^{-11}
Gravity on Earth = 1	Gravity on Jupiter	2.305
Gravity on Earth = 1	Gravity on Mars	0.3627 Equatorial
Gravity on Earth = 1	Gravity on Mercury	0.3648 Equatorial
Gravity on Earth = 1	Gravity on Moon	0.1652 Equatorial
Gravity on Earth = 1	Gravity on Neptune	1.323 ± 0.210 Equatorial
Gravity on Earth = 1	Gravity on Pluto	0.0225 ± 0.217 Equatorial
Gravity on Earth = 1	Gravity on Saturn	0.8800 Equatorial
Gravity on Earth = 1	Gravity on Sun	27.905 Equatorial

(continued)

Easy Science Demos & Labs:
Physics

From	To	× By
Gravity on Earth = 1	Gravity on Uranus	0.9554 ± 0.168 Equatorial
Gravity on Earth = 1	Gravity on Venus	0.9049 Equatorial
Hectares	Acres	2.4710538
Hectares	Square feet	107,639.10
Hectares	Square meters	10,000
Hectares	Square miles	0.0038610216
Hectares	Square rods	395.36861
Hectograms	Pounds (apoth. or troy)	0.26792289
Hectograms	Pounds (avdp.)	0.22046226
Hectoliters	Cubic centimeters	1.00028×10^5
Hectoliters	Cubic feet	3.531566
Hectoliters	Gallons (US liquid)	26.41794
Hectoliters	Ounces (US fluid)	3,381.497
Hectoliters	Pecks (US)	11.35136
Hectometers	Feet	328.08399
Hectometers	Rods	19.883878
Hectometers	Yards	109.3613
Horsepower	Horsepower (electric)	0.999598
Horsepower	Horsepower (metric)	1.01387
Horsepower	Kilowatts	0.745700
Horsepower	Kilowatts (International)	0.745577
Horsepower-hours	Kilowatts-hours	0.745700
Horsepower-hours	Watt-hours	745.700
Hours (mean solar)	Days (mean solar)	0.0416666
Hours (mean solar)	Days (sidereal)	0.041780746
Hours (mean solar)	Hours (sidereal)	1.00273791
Hours (mean solar)	Minutes (mean solar)	60
Hours (mean solar)	Minutes (sidereal)	60.164275
Hours (mean solar)	Seconds (mean solar)	3,600
Hours (mean solar)	Seconds (sidereal)	3,609.8565
Hours (mean solar)	Weeks (mean calendar)	0.0059523809
Hours (sidereal)	Days (mean solar)	0.41552899
Hours (sidereal)	Days (sidereal)	0.0416666
Hours (sidereal)	Hours (mean solar)	0.99726957
Hours (sidereal)	Minutes (mean solar)	59.836174
Hours (sidereal)	Minutes (sidereal)	60
Inches	Ångström units	2.54×10^8
Inches	Centimeters	2.54
Inches	Cubits	0.055555

(continued)

14. Units: Conversions and Constants *(continued)*

From	To	$\times$ **By**
Inches	Fathoms	0.013888
Inches	Feet	0.083333
Inches	Meters	0.0254
Inches	Mils	1,000
Inches	Yards	0.027777
Kilograms	Drams (apoth. or troy)	257.20597
Kilograms	Drams (avdp.)	564.38339
Kilograms	Dynes	980,665
Kilograms	Grains	15,432.358
Kilograms	Hundredweights (long)	0.019684131
Kilograms	Hundredweights (short)	0.022046226
Kilograms	Ounces (apoth. or troy)	32.150737
Kilograms	Ounces (avdp.)	35.273962
Kilograms	Pennyweights	643.01493
Kilograms	Pounds (apoth. or troy)	2.6792289
Kilograms	Pounds (avdp.)	2.2046226
Kilograms	Quarters (US long)	0.0039368261
Kilograms	Scruples (apoth.)	771.61792
Kilograms	Tons (long)	0.00098420653
Kilograms	Tons (metric)	0.001
Kilograms	Tons (short)	0.0011023113
Kilograms/cubic meter	Grams/cubic centimeter	0.001
Kilograms/cubic meter	Pounds/cubic foot	0.062427961
Kilograms/cubic meter	Pounds/cubic inch	3.6127292×10^{-5}
Kiloliters	Cubic centimeters	1×10^{6}
Kiloliters	Cubic feet	35.31566
Kiloliters	Cubic inches	61,025.45
Kiloliters	Cubic meters	1.000028
Kiloliters	Cubic yards	1.307987
Kiloliters	Gallons (US dry)	27.0271
Kiloliters	Gallons (US liquid)	264.1794
Kilometers	Astronomical units	6.68878×10^{-9}
Kilometers	Feet	3,280.8399
Kilometers	Light-years	1.05702×10^{-13}
Kilometers	Miles (nautical International)	0.53995680
Kilometers	Miles (statute)	0.62137119
Kilometers	Rods	198.83878
Kilometers	Yards	1,093.6133
Kilometers/hour	Centimeters/second	27.7777

(continued)

From	To	× By
Kilometers/hour	Feet/hour	3,280.8399
Kilometers/hour	Feet/minute	54.680665
Kilometers/hour	Knots (International)	0.53995680
Kilometers/hour	Meters/second	0.277777
Kilometers/hour	Miles (statute)/hour	0.62137119
Kilometers/minute	Centimeters/second	1,666.666
Kilometers/minute	Feet/minute	3,280.8399
Kilometers/minute	Kilometers/hour	60
Kilometers/minute	Knots (International)	32.397408
Kilometers/minute	Miles/hour	37.282272
Kilometers/minute	Miles/minute	0.62137119
Kilowatt-hours	Joules	3.6×10^6
Light, velocity of	Kilometers/second ± 1.1	299,792.4562 (meters/second 100 × more accurate)
Light, velocity of	Meters/second ± 0.33 ppm	2.9979250×10^8
Light, velocity of	Centimeters/second ± 0.33 ppm	2.9979250×10^{10}
Light-years	Astronomical units	63,279.5
Light-years	Kilometers	9.46055×10^{12}
Light-years	Miles (statute)	5.87851×10^{12}
Liters	Bushels (US)	0.02837839
Liters	Cubic centimeters	1,000
Liters	Cubic feet	0.03531566
Liters	Cubic inches	61.02545
Liters	Cubic meters	0.001
Liters	Cubic yards	0.001307987
Liters	Drams (US fluid)	270.5198
Liters	Gallons (US dry)	0.2270271
Liters	Gallons (US liquid)	0.2641794
Liters	Gills (US)	8.453742
Liters	Hogsheads	0.004193325
Liters	Minims (US)	16,231.19
Liters	Ounces (US fluid)	33.81497
Liters	Pecks (US)	0.1135136
Liters	Pints (US dry)	1.816217
Liters	Pints (US liquid)	2.113436
Liters	Quarts (US dry)	0.9081084
Liters	Quarts (US liquid)	1.056718
Liters/minute	Cubic feet/minute	0.03531566
Liters/minute	Cubic feet/second	0.0005885943

(continued)

14. Units: Conversions and Constants (continued)

From	To	× By
Liters/minute	Gallons (US liquid)/minute	0.2641794
Liters/second	Cubic feet/minute	2.118939
Liters/second	Cubic feet/second	0.03531566
Liters/second	Cubic yards/minute	0.07847923
Liters/second	Gallons (US liquid)/minute	15.85077
Liters/second	Gallons (US liquid)/second	0.2641794
Lumens	Candle power	0.079577472
Meters	Ångström units	1×10^{10}
Meters	Fathoms	0.54680665
Meters	Feet	3.2808399
Meters	Furlongs	0.0049709695
Meters	Inches	39.370079
Meters	Megameters	1×10^{-6}
Meters	Miles (nautical International)	0.00053995680
Meters	Miles (statute)	0.00062137119
Meters	Millimicrons	1×10^{9}
Meters	Mils	39,370.079
Meters	Rods	0.19883878
Meters	Yards	1.0936133
Meters/hour	Feet/hour	3.2808399
Meters/hour	Feet/minute	0.054680665
Meters/hour	Knots (International)	0.00053995680
Meters/hour	Miles (statute)/hour	0.00062137119
Meters/minute	Centimeters/second	1.666666
Meters/minute	Feet/minute	3.2808399
Meters/minute	Feet/second	0.054680665
Meters/minute	Kilometers/hour	0.06
Meters/minute	Knots (International)	0.032397408
Meters/minute	Miles (statute)/hour	0.037282272
Meters/second	Feet/minute	196.85039
Meters/second	Feet/second	3.2808399
Meters/second	Kilometers/hour	3.6
Meters/second	Kilometers/minute	0.06
Meters/second	Miles (statute)/hour	2.2369363
Meter-candles	Lumens/square meter	1
Micrograms	Grams	1×10^{-6}
Micrograms	Milligrams	0.001
Micromicrons	Ångström units	0.01
Micromicrons	Centimeters	1×10^{-10}

(continued)

From	To	× By
Micromicrons	Inches	$3.9370079 \times 10^{-11}$
Micromicrons	Meters	1×10^{-12}
Micromicrons	Microns	1×10^{-6}
Microns	Ångström units	10,000
Microns	Centimeters	0.0001
Microns	Feet	3.2808399×10^{-6}
Microns	Inches	3.9370070×10^{-5}
Microns	Meters	1×10^{-6}
Microns	Millimeters	0.001
Microns	Millimicrons	1,000
Miles (statute)	Centimeters	160,934.4
Miles (statute)	Feet	5,280
Miles (statute)	Furlongs	8
Miles (statute)	Inches	63,360
Miles (statute)	Kilometers	1.609344
Miles (statute)	Light-years	1.70111×10^{-13}
Miles (statute)	Meters	1,600.344
Miles (statute)	Miles (nautical International)	0.86897624
Miles (statute)	Myriameters	0.1609344
Miles (statute)	Rods	320
Miles (statute)	Yards	1,760
Miles/hour	Centimeters/second	44.704
Miles/hour	Feet/hour	5,280
Miles/hour	Feet/minute	88
Miles/hour	Feet/second	1.466666
Miles/hour	Kilometers/hour	1.609344
Miles/hour	Knots (International)	0.86897624
Miles/hour	Meters/minute	26.8224
Miles/hour	Miles/minute	0.0166666
Miles/minute	Centimeters/second	2,682.24
Miles/minute	Feet/hour	316,800
Miles/minute	Feet/second	88
Miles/minute	Kilometers/minute	1.609344
Miles/minute	Knots (International)	52.138574
Miles/minute	Meters/minute	1,609.344
Miles/minute	Miles/hour	60
Milligrams	Carats (1877)	0.004871
Milligrams	Carats (metric)	0.005
Milligrams	Drams (apoth. or troy)	0.00025720597

(continued)

14. Units: Conversions and Constants *(continued)*

From	To	× By
Milligrams	Drams (advp.)	0.00056438339
Milligrams	Grains	0.015432358
Milligrams	Grams	0.001
Milligrams	Ounces (apoth. or troy)	3.2150737×10^{-5}
Milligrams	Ounces (avdp.)	3.5273962×10^{-5}
Milligrams	Pounds (apoth. or troy)	2.6792289×10^{-6}
Milligrams	Pounds (avdp.)	2.2046226×10^{-6}
Milligrams/liter	Grains/gallon (US)	0.05841620
Milligrams/liter	Grams/liter	0.001
Milligrams/liter	Parts/million	1; solvent density = 1
Milligrams/liter	Pounds/cubic foot	6.242621×10^{-5}
Milligrams/millimeter	Dynes/centimeter	9.80665
Milliliters	Cubic centimeters	1
Milliliters	Cubic inches	0.06102545
Milliliters	Drams (US fluid)	0.2705198
Milliliters	Gills (US)	0.008453742
Milliliters	Minims (US)	16.23119
Milliliters	Ounces (US fluid)	0.03381497
Milliliters	Pints (US liquid)	0.002113436
Millimeters	Ångström units	1×10^{7}
Millimeters	Centimeters	0.1
Millimeters	Decimeters	0.01
Millimeters	Dekameters	0.0001
Millimeters	Feet	0.0032808399
Millimeters	Inches	0.039370079
Millimeters	Meters	0.001
Millimeters	Microns	1,000
Millimeters	Mils	39.370079
Millimicrons	Ångström units	10
Millimicrons	Centimeters	1×10^{-7}
Millimicrons	Inches	3.9370079×10^{-8}
Millimicrons	Microns	0.001
Millimicrons	Millimeters	1×10^{-6}
Minutes (angular)	Degrees	0.0166666
Minutes (angular)	Quadrants	0.000185185
Minutes (angular)	Radians	0.00029088821
Minutes (angular)	Seconds (angular)	60
Minutes (mean solar)	Days (mean solar)	0.0006944444
Minutes (mean solar)	Days (sidereal)	0.00069634577

(continued)

From	To	× By
Minutes (mean solar)	Hours (mean solar)	0.0166666
Minutes (mean solar)	Hours (sidereal)	0.016732298
Minutes (mean solar)	Minutes (sidereal)	1.00273791
Minutes (sidereal)	Days (mean solar)	0.00069254831
Minutes (sidereal)	Minutes (mean solar)	0.99726957
Minutes (sidereal)	Months (mean calendar)	2.2768712×10^{-5}
Minutes (sidereal)	Seconds (sidereal)	60
Minutes/centimeter	Radians/centimeter	0.00029088821
Months (lunar)	Days (mean solar)	29.530588
Months (lunar)	Hours (mean solar)	708.73411
Months (lunar)	Minutes (mean solar)	42,524.047
Months (lunar)	Seconds (mean solar)	2.5514428×10^{-5}
Months (lunar)	Weeks (mean calendar)	4.2186554
Months (mean calendar)	Days (mean solar)	30.416666
Months (mean calendar)	Hours (mean solar)	730
Months (mean calendar)	Months (lunar)	1.0300055
Months (mean calendar)	Weeks (mean calendar)	4.3452381
Months (mean calendar)	Years (calendar)	0.08333333
Months (mean calendar)	Years (sidereal)	0.083274845
Months (mean calendar)	Years (tropical)	0.083278075
Myriagrams	Pounds (avdp.)	22.046226
Ounces (avdp.)	Drams (apoth. or troy)	7.291666
Ounces (avdp.)	Drams (avdp.)	16
Ounces (avdp.)	Grains	437.5
Ounces (avdp.)	Grams	28.349
Ounces (avdp.)	Ounces (apoth. or troy)	0.9114583
Ounces (avdp.)	Pounds (apoth. or troy)	0.075954861
Ounces (avdp.)	Pounds (avdp.)	0.0625
Ounces (US fluid)	Cubic centimeters	29.573730
Ounces (US fluid)	Cubic inches	1.8046875
Ounces (US fluid)	Cubic meters	2.9573730×10^{-5}
Ounces (US fluid)	Drams (US fluid)	8
Ounces (US fluid)	Gallons (US dry)	0.0067138047
Ounces (US fluid)	Gallons (US liquid)	0.0078125
Ounces (US fluid)	Gills (US)	0.25
Ounces (US fluid)	Liters	0.029572702
Ounces (US fluid)	Pints (US liquid)	0.0625
Ounces (US fluid)	Quarts (US liquid)	0.03125
Ounces/square inch	Dynes/square centimeter	4309.22

(*continued*)

From	To	× By
Ounces/square inch	Grams/square centimeter	4.3941849
Ounces/square inch	Pounds/square foot	9
Ounces/square inch	Pounds/square inch	0.0625
Parts/million	Grains/gallon (US)	0.05841620
Parts/million	Grams/liter	0.001
Parts/million	Milligrams/liter	1
Pints (US dry)	Bushels (US)	0.015625
Pints (US dry)	Cubic centimeters	550.61047
Pints (US dry)	Cubic inches	33.6003125
Pints (US dry)	Gallons (US dry)	0.125
Pints (US dry)	Gallons (US liquid)	0.14545590
Pints (US dry)	Liters	0.5505951
Pints (US dry)	Pecks (US)	0.0625
Pints (US dry)	Quarts (US dry)	0.5
Pints (US liquid)	Cubic centimeters	473.17647
Pints (US liquid)	Cubic feet	0.016710069
Pints (US liquid)	Cubic inches	28.875
Pints (US liquid)	Cubic yards	0.00061889146
Pints (US liquid)	Drama (US fluid)	128
Pints (US liquid)	Gallons (US liquid)	0.125
Pints (US liquid)	Gills (US)	4
Pints (US liquid)	Liters	0.4731632
Pints (US liquid)	Milliliters	473.1632
Pints (US liquid)	Minims (US)	7,680
Pints (US liquid)	Ounces (US fluid)	16
Pints (US liquid)	Quarts (US liquid)	0.5
Planck's constant	Erg-seconds	6.6255×10^{-27}
Planck's constant	Joule-seconds	6.6255×10^{-34}
Planck's constant	Joule-seconds/Avog. No. (chem.)	3.9905×10^{-10}
Pounds (apoth. or troy)	Drams (apoth. or troy)	96
Pounds (apoth. or troy)	Drams (avdp.)	210.65143
Pounds (apoth. or troy)	Grains	5,780
Pounds (apoth. or troy)	Grams	373.24172
Pounds (apoth. or troy)	Kilograms	0.37324172
Pounds (apoth. or troy)	Ounces (apoth. or troy)	12
Pounds (apoth. or troy)	Ounces (avdp.)	13.165714
Pounds (apoth. or troy)	Pounds (avdp.)	0.8228571
Pounds (avdp.)	Drams (apoth. or troy)	116.6686
Pounds (avdp.)	Drams (avdp.)	256

(continued)

14. Units: Conversions and Constants *(continued)*

From	To	× **By**
Pounds (avdp.)	Grains	7,000
Pounds (avdp.)	Grams	453.59237
Pounds (avdp.)	Kilograms	0.45359237
Pounds (avdp.)	Ounces (apoth. or troy)	14.593333
Pounds (avdp.)	Ounces (avdp.)	16
Pounds (avdp.)	Pounds (apoth. or troy)	1.215277
Pounds (avdp.)	Scruples (apoth.)	350
Pounds (avdp.)	Tons (long)	0.00044642857
Pounds (avdp.)	Tons (metric)	0.00045359237
Pounds (avdp.)	Tons (short)	0.0005
Pounds/cubic foot	Grams/cubic centimeter	0.016018463
Pounds/cubic foot	Kilograms/cubic meter	16.018463
Pounds/cubic inch	Grams/cubic centimeter	27.679905
Pounds/cubic inch	Grams/liter	27.68068
Pounds/cubic inch	Kilograms/cubic meter	27,679.005
Pounds/gallon (US liquid)	Grams/cubic centimeter	0.11982643
Pounds/gallon (US liquid)	Pounds/cubic foot	7.4805195
Pounds/inch	Grams/centimeter	178.57967
Pounds/inch	Grams/foot	5,443.1084
Pounds/inch	Grams/inch	453.59237
Pounds/inch	Ounces/centimeter	6.2992
Pounds/inch	Ounces/inch	16
Pounds/inch	Pounds/meter	39.370079
Pounds/minute	Kilograms/hour	27.2155422
Pounds/minute	Kilograms/minute	0.45359237
Pounds on Earth = 1	Pounds on Jupiter	2.529 Equatorial
Pounds on Earth = 1	Pounds on Mars	0.3627 Equatorial
Pounds on Earth = 1	Pounds on Mercury	0.3648 Equatorial
Pounds on Earth = 1	Pounds on Moon	0.1652 Equatorial
Pounds on Earth = 1	Pounds on Neptune	1.323 ± 0.210 Equatorial
Pounds on Earth = 1	Pounds on Pluto	0.0225 ± 0.217 Equatorial
Pounds on Earth = 1	Pounds on Saturn	0.8800 Equatorial
Pounds on Earth = 1	Pounds on Sun	27.905 Equatorial
Pounds on Earth = 1	Pounds on Uranus	0.9554 ± 0.168 Equatorial
Pounds on Earth = 1	Pounds on Venus	0.9049 Equatorial
Pounds/square foot	Atmospheres	0.000472541
Pounds/square foot	Bars	0.000478803
Pounds/square foot	Centimeter of Hg (0°C)	0.0359131
Pounds/square foot	Dynes/square centimeter	478.803

(continued)

14. Units: Conversions and Constants *(continued)*

From	To	× By
Pounds/square foot	Feet of air (1 atm. 60°F)	13.096
Pounds/square foot	Grams/square centimeter	0.48824276
Pounds/square foot	Kilograms/square meter	4.8824276
Pounds/square foot	Millimeters of Hg (0°C)	0.369131
Pounds/square inch	Atmospheres	0.0680460
Pounds/square inch	Bars	0.0689476
Pounds/square inch	Dynes/square centimeter	68,947.6
Pounds/square inch	Grams/square centimeter	70.306958
Pounds/square inch	Kilograms/square centimeter	0.070306958
Pounds/square inch	Millimeters of Hg (0°C)	51.7149
Quarts (US dry)	Bushels (US)	0.03125
Quarts (US dry)	Cubic centimeters	1,101.2209
Quarts (US dry)	Cubic feet	0.038889251
Quarts (US dry)	Cubic inches	67.200625
Quarts (US dry)	Gallons (US dry)	0.25
Quarts (US dry)	Gallons (US liquid)	0.29091180
Quarts (US dry)	Liters	1.1011901
Quarts (US dry)	Pecks (US)	0.125
Quarts (US dry)	Pints (US dry)	2
Quarts (US liquid)	Cubic centimeters	946.35295
Quarts (US liquid)	Cubic feet	0.033420136
Quarts (US liquid)	Cubic inches	57.75
Quarts (US liquid)	Drams (US fluid)	256
Quarts (US liquid)	Gallons (US dry)	0.21484175
Quarts (US liquid)	Gallons (US liquid)	0.25
Quarts (US liquid)	Gills (US)	8
Quarts (US liquid)	Liters	0.9463264
Quarts (US liquid)	Ounces (US fluid)	32
Quarts (US liquid)	Pints (US liquid)	2
Quarts (US liquid)	Quarts (US dry)	0.8593670
Quintals (metric)	Grams	100,000
Quintals (metric)	Hundredweights (long)	1.9684131
Quintals (metric)	Kilograms	100
Quintals (metric)	Pounds (avdp.)	220.46226
Radians	Circumferences	0.15915494
Radians	Degrees	57.295779
Radians	Minutes	3,437.7468
Radians	Quadrants	0.63661977
Radians	Revolutions	0.15915494

(continued)

14. Units: Conversions and Constants (continued)

From	To	× By
Revolutions	Degrees	360
Revolutions	Grades	400
Revolutions	Quadrants	4
Revolutions	Radians	6.2831853
Seconds (angular)	Degrees	0.000277777
Seconds (angular)	Minutes	0.0166666
Seconds (angular)	Radians	4.8481368×10^{-6}
Seconds (mean solar)	Days (mean solar)	1.1574074×10^{-5}
Seconds (mean solar)	Days (sidereal)	1.1605763×10^{-5}
Seconds (mean solar)	Hours (mean solar)	0.0002777777
Seconds (mean solar)	Hours (sidereal)	0.00027853831
Seconds (mean solar)	Minutes (mean solar)	0.0166666
Seconds (mean solar)	Minutes (sidereal)	0.016712298
Seconds (mean solar)	Seconds (sidereal)	1.00273791
Seconds (sidereal)	Days (mean solar)	1.1542472×10^{-5}
Seconds (sidereal)	Days (sidereal)	1.1574074×10^{-5}
Seconds (sidereal)	Hours (mean solar)	0.00027701932
Seconds (sidereal)	Hours (sidereal)	0.000277777
Seconds (sidereal)	Minutes (mean solar)	0.016621159
Seconds (sidereal)	Minutes (sidereal)	0.0166666
Seconds (sidereal)	Seconds (mean solar)	0.09726957
Square centimeters	Square decimeters	0.01
Square centimeters	Square feet	0.0010763910
Square centimeters	Square inches	0.15500031
Square centimeters	Square meters	0.0001
Square centimeters	Square millimeters	100
Square centimeters	Square miles	1.5500031×10^{5}
Square centimeters	Square yards	0.00011959900
Square decimeters	Square centimeters	100
Square decimeters	Square inches	15.500031
Square dekameters	Acres	0.024710538
Square dekameters	Ares	1
Square dekameters	Square meters	100
Square dekameters	Square yards	119.59900
Square feet	Acres	2.295684×10^{-5}
Square feet	Ares	0.0009290304
Square feet	Square centimeters	929.0304
Square feet	Square inches	144
Square feet	Square meters	0.09290304

(continued)

14. Units: Conversions and Constants *(continued)*

From	To	× By
Square feet	Square miles	3.5870064×10^{-8}
Square feet	Square yards	0.111111
Square hectometers	Square meters	10,000
Square inches	Square centimeters	6.4516
Square inches	Square decimeters	0.064516
Square inches	Square feet	0.0069444
Square inches	Square meters	0.00064516
Square inches	Square miles	$2.4909767 \times 10^{-10}$
Square inches	Square millimeters	645.16
Square inches	Square mils	1×10^{-6}
Square kilometers	Acres	247.10538
Square kilometers	Square feet	1.0763010×10^{7}
Square kilometers	Square inches	1.5500031×10^{9}
Square kilometers	Square meters	1×10^{6}
Square kilometers	Square miles	0.38610216
Square kilometers	Square yards	1.1959900×10^{6}
Square meters	Acres	0.00024710538
Square meters	Ares	0.01
Square meters	Hectares	0.0001
Square meters	Square centimeters	10,000
Square meters	Square feet	10.763910
Square meters	Square inches	1,550.0031
Square meters	Square kilometers	1×10^{-6}
Square meters	Square miles	3.8610218×10^{-7}
Square meters	Square millimeters	1×10^{6}
Square meters	Square yards	1.1959900
Square miles	Acres	640
Square miles	Hectares	258.99881
Square miles	Square feet	2.7878288×10^{7}
Square miles	Square kilometers	2.5899881
Square miles	Square meters	2.5899881×10^{6}
Square miles	Square rods	102,400
Square miles	Square yards	3.0976×10^{6}
Square millimeters	Square centimeters	0.01
Square millimeters	Square inches	0.0015500031
Square millimeters	Square meters	1×10^{-6}
Square yards	Acres	0.00020661157
Square yards	Ares	0.0083612736
Square yards	Hectares	8.3612736×10^{-5}

(continued)

From	To	× By
Square yards	Square centimeters	8,361.2736
Square yards	Square feet	9
Square yards	Square inches	1,296
Square yards	Square meters	0.83612736
Square yards	Square miles	$3.228305785 \times 10^{-7}$
Tons (long)	Kilograms	1,016.0469
Tons (long)	Ounces (avdp.)	35,840
Tons (long)	Pounds (apoth. or troy)	2,722.22
Tons (long)	Pounds (avdp.)	2,240
Tons (long)	Tons (metric)	1.0160469
Tons (long)	Tons (short)	1.12
Tons (metric)	Dynes	9.80665×10^{8}
Tons (metric)	Grams	1×10^{6}
Tons (metric)	Kilograms	1,000
Tons (metric)	Ounces (avdp.)	35,273.962
Tons (metric)	Pounds (apoth. or troy)	2,679.2289
Tons (metric)	Pounds (avdp.)	2,204.6226
Tons (metric)	Tons (long)	0.98420653
Tons (metric)	Tons (short)	1.1023113
Tons (short)	Kilograms	907.18474
Tons (short)	Ounces (avdp.)	32,000
Tons (short)	Pounds (apoth. or troy)	2,430.555
Tons (short)	Pounds (avdp.)	2,000
Tons (short)	Tons (long)	0.89285714
Tons (short)	Tons (metric)	0.90718474
Velocity of light	Centimeters/second ± 0.33 ppm	$2.9979250 \, (10) \times 10^{10}$
Velocity of light	Meters/second ± 0.33 ppm	$2.9979250 \, (10) \times 10^{8}$
Velocity of light (100 x more accurate)	Kilometers/second ± 1.1 meter/second	$2.997924562 \times 10^{5}$
Volts	Mks. (r or nr) units	1
Volts (International)	Volts	1.000330
Volts-seconds	Maxwells	1×10^{8}
Watts	Kilowatts	0.001
Watts (International)	Watts	1.000165
Weeks (mean calendar)	Days (mean solar)	7
Weeks (mean calendar)	Days (sidereal)	7.0191654
Weeks (mean calendar)	Hours (mean solar)	168
Weeks (mean calendar)	Hours (sidereal)	168.45997
Weeks (mean calendar)	Minutes (mean solar)	10,080

(continued)

14. Units: Conversions and Constants (continued)

From	To	× By
Weeks (mean calendar)	Minutes (sidereal)	10,107.598
Weeks (mean calendar)	Months (lunar)	0.23704235
Weeks (mean calendar)	Months (mean calendar)	0.23013699
Weeks (mean calendar)	Years (calendar)	0.019178082
Weeks (mean calendar)	Years (sidereal)	0.019164622
Weeks (mean calendar)	Years (tropical)	0.019165365
Yards	Centimeters	91.44
Yards	Cubits	2
Yards	Fathoms	0.5
Yards	Feet	3
Yards	Furlongs	0.00454545
Yards	Inches	36
Yards	Meters	0.9144
Yards	Rods	0.181818
Yards	Spans	4
Years (calendar)	Days (mean solar)	365
Years (calendar)	Hours (mean solar)	8,760
Years (calendar)	Minutes (mean solar)	525,600
Years (calendar)	Months (lunar)	12.360065
Years (calendar)	Months (mean calendar)	12
Years (calendar)	Seconds (mean solar)	3.1536×10^7
Years (calendar)	Weeks (mean calendar)	52.142857
Years (calendar)	Years (sidereal)	0.99929814
Years (calendar)	Years (tropical)	0.99933690
Years (leap)	Days (mean solar)	366
Years (sidereal)	Days (mean solar)	365.25636
Years (sidereal)	Days (sidereal)	366.25640
Years (sidereal)	Years (calendar)	1.0007024
Years (sidereal)	Years (tropical)	1.0000388
Years (tropical)	Days (mean solar)	365.24219
Years (tropical)	Days (sidereal)	366.24219
Years (tropical)	Hours (mean solar)	8,765.8126
Years (tropical)	Hours (sidereal)	8,789.8126
Years (tropical)	Months (mean calendar)	12.007963
Years (tropical)	Seconds (mean solar)	3.1556926×10^7
Years (tropical)	Seconds (sidereal)	3.1643326×10^7
Years (tropical)	Weeks (mean calendar)	52.177456
Years (tropical)	Years (calendar)	1.0006635
Years (tropical)	Years (sidereal)	0.99996121

A

absolute frame of reference: the now-outdated idea that motion is not relative; that velocity can be measured specifically

acceleration: the rate at which an object's velocity changes with time; the change may be in speed, or direction, or both.

acceleration due to gravity: the acceleration of a free-falling object; its value near the earth's surface is 9.8 m/s/s.

amplitude: for a wave or a vibration, the maximum displacement on either side of the equilibrium, or midpoint of the wave

angle of incidence: the angle at which light strikes a surface

angle of reflection: the angle at which light is reflected from a surface; angles of incidence and reflection are always the same, hence, the equation, "The angle of incidence is equal to the angle of reflection."

Archimedes' principle: An immersed object, submerged or floating, is buoyed up by a force equal to the weight of fluid displaced.

B

Bernoulli's principle: Pressure of a fluid on a surface decreases as the fluid's velocity relative to the surface increases.

block and tackle: a series of pulleys that reduces the effort needed to lift a weight

Brownian motion: random movement of microscopic particles suspended in liquids or gases resulting from the impact of molecules of the fluid surrounding the particles

buoyancy: the degree to which an object floats in a liquid

C

calorie: energy required to raise the temperature of one gram of water one degree C

cathode ray oscilloscope (CRO): a scientific instrument that is used to read fluctuations in an electrical quantity as a sine wave

closed system: a system that has no energy release or input, and in which entropy increases; the universe is believed to be a closed system.

coherent: Wave fronts that remain in phase; a laser is an example of a coherent light source.

compression: the region of increased pressure in a longitudinal wave

condensation: the change of state from gas to liquid

conductor: material that conducts heat or electricity

conservation of momentum: principle that the amount of momentum within a system remains constant, even when the system undergoes change

constant: a number never known to vary, or a section of an experiment kept the same while other things change to test for a particular principle

constructive interference: superimposition of waves in which the waves reinforce one another

cosmic ray: one of a variety of high-speed particles that travels through the universe, and has its origins in a supernova

(continued)

critical angle: the angle at which light no longer is refracted when it meets a surface, but instead is totally internally reflected

current: electron flow

D

decibel: a unit of measurement for the loudness of sound

density: a measurement of the number of molecules in a given space; density is calculated by dividing mass by volume.

destructive interference: superimposition of waves in which the waves cancel one another out

diffraction grating: a series of closely spaced parallel slits used to separate colors of light by interference

dipole: any object with two poles, such as a magnet

displacement: the volume or weight of a fluid (as water) displaced by a floating body (as a ship) of equal weight

drag: something that retards motion or action

E

efficiency: ratio of result to effort, output to input, involving energy

effort: the measurable quantity of work done to achieve a specific aim

elastic collision: a collision in which no energy is transformed to heat

electromagnet: a magnet that derives its magnetism from an active electric field; when the field is removed or switched off, the magnetism fails.

electromagnetic spectrum: the range of frequencies over which electromagnetic radiation can be propogated; the lowest frequencies are associated with radio waves, microwaves have a higher frequency, then infrared, visible light, ultraviolet, X rays and gamma rays.

electromotive force (EMF): any force that gives rise to an electric current; a battery or a generator is a source for EMF.

entropy: a measure of the degree of disorder in a system

equilibrium: the state of an object when not acted upon by a net force

evaporation: the change of state at the surface of a liquid as it becomes vapor

F

Fermat's principle of least time: Light will take the path that requires the least time when it goes from one place to another.

fluorescence: the property of absorbing radiation at one frequency, followed by its re-emission at a lower frequency

force: any influence that can cause an object to be accelerated, measured in newtons

force lines: generally invisible lines that extend from one pole to another, on a dipole, such as a magnet

free fall: motion of an object under the influence of gravitational pull only

frequency: number of waves that pass a particular point per unit time

friction: forces resisting motion

fulcrum: the pivot point of a lever

(continued)

G

gamma rays: high-frequency electro-magnetic radiation emitted by the nuclei of radioactive atoms

gas: state of matter beyond the liquid state, wherein molecules fill whatever space is available to them, taking no definite shape

gravity: attraction between objects due to mass

H

heat of fusion: the amount of heat energy needed to change the state of matter

hertz (Hz): unit of frequency in sound

I

inclined plane: a simple machine, such as a ramp, that decreases the effort necessary to do work, although increasing the time necessary

incoherent: wave fronts that do not remain in phase; an example of incoherent light is a flashlight, or sunlight.

inelastic collision: a collision in which some of the energy and momentum is transformed to heat

inertia: the fundamental property of inert matter tending to resist change in its state of motion

infrared rays: part of the electro-magnetic spectrum, slightly less energetic than visible light, and detectable as heat

insulator: any material through which charge strongly resists flow when electrical force is applied

interference: superimposition of waves, producing regions of reinforcement (constructive interference) and cancellation (destructive interference)

K

kinetic energy: energy of motion

kinetic molecular theory: principle that changes in states of matter are caused by motion of molecules contained within that matter

L

laser (light amplification by stimulated emission of radiation): an optical instrument that produces a beam of coherent light

law of falling bodies: principle that in a vacuum, two objects will fall at the same rate of acceleration, regardless of their mass

lens: a piece of glass or other transparent material that can focus light

lever: a simple machine, made of a bar that pivots around a fixed point, called a fulcrum

linear expansivity: the amount of linear expansion when heat is applied

liquid: the state of matter between solid and gas, in which the matter possesses a definite volume but no definite shape

longitudinal wave: a wave in which the individual particles of the medium vibrate back and forth in the direction the wave travels; sound is an example.

Lorentz transformation: an equation that demonstrates the increasing mass, decreasing length, or dilation of time that are predicted by the Special Theory of Relativity

(continued)

M

magnet: a piece of ferrous or cobalt in which molecules and atoms are lined up in a particular orientation, and which can be used to attract other ferrous-bearing objects

magnetic field: the region of altered space that will interact with the magnetic properties of a magnet

manometer: a scientific device for measurement of pressure

mass: the quantity of matter in an object; the measurement of the inertia of an object

mechanical advantage: a calculation of how many times a machine multiplies the effort used

microwaves: part of the electromagnetic spectrum, microwaves have a higher frequency than radio waves but a lower frequency than infrared waves.

moment: measure of a force's ability to rotate an object

momentum: the product of the mass of an object and its velocity

monochromatic: in light, one-frequency light beam, usually coherent, such as a laser beam

N

natural frequency: the particular frequency at which an object, such as a tuning fork, vibrates

newton: the SI (Systeme Internationale) unit of measure of force; a force of 1 newton accelerates a mass of 1 kilogram 1 meter per second per second.

Newton's laws of motion:

Law 1: Every object continues in its state of rest or of uniform motion in a straight line unless it is compelled to change that state by forces impressed upon it.

Law 2: The acceleration of an object is directly proportional to the net force acting on the object and is inversely proportional to the mass of the object ($F = ma$).

Law 3: To every action force, there is an equal and opposite reaction force.

O

Ohm's law: The current in a circuit varies in direct proportion to the potential difference or EMF, and in inverse proportion to resistance. Current = voltage/resistance, or $I = V/R$.

P

parallel circuit: an electric circuit with two or more resistances arranged in branching paths in such a way that any single one completes the circuit independently of the other

phase: in light, when wave fronts arrive at a destination without spreading out; a laser provides coherent light that arrives in phase; a flashlight provides incoherent light that arrives out of phase.

photon: a particle of light

physical change: a change that occurs to matter without changing the chemical make-up of the matter

pole: in a magnet, the specific location from which force lines emanate

(continued)

potential energy: the stored energy that an object possesses because of its position relative to other objects; e.g., a marble at the top of a hill has a greater potential energy than an identical marble at the bottom of a hill.

pressure: the ratio of the amount of force to the area over which the force is distributed: Pressure = force/area.

prism: a usually triangular piece of material, such as glass, that separates incident light by refraction into its component colors

property: a statement about a particular attribute or attributes of a material

pulley: a simple machine; a wheel that turns on an axle and can change the direction of force to increase efficiency

Q

quantum theory: The physical theory based on the idea that energy is radiated in discrete units called quanta; in light, energy is radiated in photons; further, that all particle matter also has a wave nature

R

radiation: the transfer of energy by means of the electromagnetic spectrum or high-speed particles

radio waves: energy with the lowest known frequency of the electromagnetic spectrum

reflection: the bouncing of light rays from a surface, such that the angle of incidence is equal to the angle of reflection

refraction: the bending light when it passes from one transparent medium to another, caused by a difference in the speed of light in those media

relativistic effects: apparent effects that take place to an observer as the velocity of an object increases toward the speed of light; these are an increase in the object's mass, a decrease in length, and a dilation—or slowing—of time

resistance: in electricity, an object or condition that slows down a current through a circuit; light bulbs are resistors, for example.

resistivity: a property of matter, resistance in any given material to conduct an electric current

resonance: the setting up of vibrations in an object at its natural vibration frequency by a vibrating force or wave having the same frequency

S

scalar: any quantity without direction, but with a magnitude, such as speed, mass, volume, etc.

schlieren: stringers of coloring that have a lower density than the surrounding solution, and seem to drop to the bottom of a container; this is a property of some materials.

second law of thermodynamics: Heat cannot be transferred from a colder body to a hotter body without work being done by an outside agent.

series circuit: an electric current with devices that have resistance arranged so that the same electric current flows through all of them

short circuit: A circuit in which there are no resistances and in which the current passes back to the source in a very short period of time; these can be dangerous, since the rapid nature of the

(continued)

process causes the source to heat dramatically.

S.I. (Systeme Internationale): modern system of definitions and metric notation, used in science worldwide

simple circuit: a circuit with one resistor

simple machine: any of several machines designed to increase efficiency and reduce effort, including the inclined plane, the pulley, the wheel and axle, the levers, the wedge, and the screw

slope: in a graph, the relative steepness, or positive and negative nature, of a straight line graph following the equation $y = mx + b$, where y and x are positions on the axes, b is the y intercept, and m is the slope

solid: the state of matter characterized by definite volume and shape

solubility: the ability to be dissolved in a solution

solvency: the ability to dissolve another substance

special relativity: a formulation of the consequences of the absence of an absolute frame of reference; the theory has two postulates: 1) All laws of nature are the same in all uniformly moving frames of reference, i.e., those moving with constant velocity, as opposed to accelerating frames; and 2) The speed of light in a vacuum will have the same value, regardless of the motion of the observer.

specific gravity: the density of a material without its units; this is accomplished by dividing the density of a material by that of water.

specific heat: the quantity of heat per unit mass required to raise the temperature either 1 degree C or 1 degree K; it is measured in calories for Celsius measurements, and in joules for Kelvin measurements.

speed: the time rate at which distance is covered by a moving object

static electricity: electricity that is collected by rubbing certain surfaces together, causing electrons to flow from one to the other, charging an object that is typically neutrally charged

streamline: to change the shape of an object in order to offer less resistance to movement in air or in water

superconductor: a material that is a perfect conductor with zero resistance to the flow of electrical charge

T

thermal conductivity: a property of matter, the ability of a material to conduct heat

thermodynamics: the physics of the interrelationships between heat and other forms of energy, characterized by two laws. The first law: A restatement of the law of conservation of energy as it applies to systems involving change of temperature. Whenever heat is added to a system, it transforms to an equal amount of some other form of energy. The second law: Heat cannot be transferred from a colder body to a hotter body without work being done by an outside agent. A consequence of the second law is the principle of entropy.

(continued)

torque: the product of force and lever arm distance, which tends to produce rotation

total internal reflection: the total reflection of light traveling in a medium when it is incident on the surface of a less dense medium at an angle greater than the critical angle

transverse wave: a wave in which the individual particles of a medium vibrate from side to side across the direction in which the wave travels; light is a transverse wave.

U

ultraviolet rays: Part of the electromagnetic spectrum, ultraviolet rays have a greater frequency than visible light, but less than X rays.

V

Van de Graaf generator: a laboratory device for generating static electricity

vector: a quantity that has both magnitude and direction, such as velocity, force, acceleration, etc.

velocity: the speed of an object in a particular direction, a vector quantity

visible light: Part of the electromagnetic spectrum, visible light has frequencies greater than infrared and lower than ultraviolet radiation.

voltage: electrical pressure, or a measure of electrical potential difference

volume: the quantity of space an object occupies

W

wavelength: the distance between successive crests in a wave

weight: the force of Earth's gravitational attraction for any object on, above, or below Earth's surface

wheel and axle: A simple machine, a wheel and axle is a mechanical device consisting of a grooved wheel turned by a cord or chain with a rigidly attached axle which reduces effort and increases efficiency.

work: the product of the force exerted and the distance through which the force moves: $W = Fd$.

X

X rays: Part of the electromagnetic spectrum, X rays have a higher frequency than ultraviolet rays and a lower frequency than gamma rays.

Y

Young's modulus: a measurement of the elasticity of a material, and of how much it can be stretched before it deforms

Young's two slit experiment: an experiment in which light was beamed through two slits in order to show interference patterns